Patrick Zoll

What It Is to Exist

Quellen und Studien zur Philosophie

Edited by
Dominik Perler and Michael Quante

Volume 149

Patrick Zoll

What It Is to Exist

The Contribution of Thomas Aquinas's View to the Contemporary Debate

DE GRUYTER

ISBN 978-3-11-099130-7
e-ISBN (PDF) 978-3-11-097987-9
e-ISBN (EPUB) 978-3-11-098012-7
ISSN 0344-8142

Library of Congress Control Number: 2022942224

Bibliographic information published by the Deutsche Nationalbibliothek
The Deutsche Nationalbibliothek lists this publication in the Deutsche Nationalbibliografie; detailed bibliographic data are available on the internet at http://dnb.dnb.de.

Printing and binding: CPI books GmbH, Leck

www.degruyter.com

Acknowledgments

This book is based on a manuscript which was accepted by the University of Innsbruck in April 2022 as fulfilling the requirements of a habilitation. The writing and completion of the manuscript has greatly benefited from three longer research periods at Saint Louis University, MO, USA. I am grateful to the members of the Jesuit communities, to the staff and students of the Catholic Studies Center, to my colleagues at the philosophy department, and to other persons who made these stays possible, warmly welcomed me, and made me feel truly at home during these times. In particular, I would like to mention: Michael Barber, Gregory Beabout, Vincent Bernabé Dávila, Justin Daffron, Chad W. Flanders, Tina Godar, John Greco, Sean Hagerty, Dan Haybron, Don Highberger, Barbara Manning, Colleen McCluskey, David Meconi, Thomas Melancon, Heidi Moore, Ronny O'Dwyer, Scott Ragland, William Regh, Sean Salai, Donald Stump, Theodore Vitali, and John Padberg. I am especially indebted to Michael and Rita Mooney whose generosity allowed me to accept a position as a postdoctoral fellow from 2019 to 2020.

Of invaluable help was Eleonore Stump's dissertation group with its varying members from 2016 to 2021. I would like to mention those with whom a daily email routine became a joyful experience of mutual support when physical meetings were no longer feasible during the first lockdowns due to the pandemic: Patrick McCaffery, Emily McCarty, Jonathan Nebel, Alexandra Romanyshyn, James Dominic Rooney, Shannon Sandridge, Hayden Stephan, Katherine Sweet, Seyma Yazici, Chong Yuan, and Yling Zhou.

I very much appreciate the generous help of colleagues and friends who read parts of the manuscript, or all of it: Jason Eberl, Joanne Falzon, Domink Finkelde, Ann Garcia, Rosemary Jermann, Ralf Klein, John Peck, James Dominic Rooney, Georg Sans, Jeremy Skrzypek, and Olivia Winiger. Their numerous comments and suggestions helped improve the argument and the prose significantly. All remaining faults and infelicities are of course only attributable to me.

I would like to thank Godehard Brüntrup, Bruno Niederbacher, and Timothy Pawl for their willingness to function as referees. Their reports were very valuable and aided me in revising my manuscript for publication. A debt of gratitude is owed to Josef Quitterer for guiding me through the habilitation procedure at the University of Innsbruck.

I am grateful to the editors of this series, Dominik Perler and Michael Quante, for accepting my manuscript for publication. Their comments and suggestions were exceptionally helpful in shaping the final form of the book. I owe

https://doi.org/10.1515/9783110979879-001

my gratitude to Marcus Böhm from De Gruyter for his untiring support regarding all technicalities.

A person whom I could have named in nearly all the regards mentioned above is Eleonore Stump. I would like to single her out because without her this book would not have seen the light of day. She introduced me to the thought of Thomas Aquinas, welcomed me in St. Louis, and provided me with constant and invaluable feedback at the different stages of this project. It was and is an honor and privilege to work with her, not only because she is an exceptional researcher but especially because of her way of teaching, which is informed by the Jesuit pedagogical ideal of *cura personalis*—'care of the whole person.' Being a Jesuit myself, it is especially inspiring to witness the potential of this ideal when it is put into practice like in her case. It is deeply consoling to see and experience how her students—and I include myself in this category—flourish as philosophers and human beings through her care for them. With deep gratitude and affection, I would like to dedicate this book to her, and I hope that I will be able to pay back at least a bit of what I owe her by trying to teach my students with the same patience, dedication, and care.

Contents

List of Abbreviations

A Works by Thomas Aquinas

DEE	De ente et essentia
DPN	De principiis naturae
DSS	De substantiis separatis
In BDH	Expositio in librum Boethii De hebdomadibus
In BDT	Expositio super librum Boethii De trinitate
In CA	Expositio super librum De causis
In DDN	Expositio super librum Dionysii De divinis nominibus
In Meta	Sententia super Metaphysicam
In Phy	Sententia super Physicam
QDA	Quaestio disputata de anima
QDP	Quaestiones disputatae de potentia
QDSC	Quaestio disputata de spiritualibus creaturis
QDUVI	Quaestio disputata de unione verbi incarnati
QDV	Quaestiones disputatae de veritate
QQ	Quaestiones quodlibetales
SCG	Summa contra gentiles
ST	Summa theologiae

B Latin Abbreviations

a.	articulus
ad.	adversus
cap.	caput
corp.	corpus
obj.	objectio
lect.	lectio
q.	quaestio

C Other Abbreviations

BMC	basic modal characterization
ECMC	existence-conditioned modal characterization

https://doi.org/10.1515/9783110979879-002

PWC	possible world characterization
MPWC	modified possible world characterization
UDO	unrestricted domain of objects
REDO	restricted domain of objects

Introduction

There is a widespread consensus among contemporary philosophers that there are entities which exist. Objects like dogs, cats, trees, puddles of water, chunks of gold, and, of course, human beings are often considered to be intuitively plausible and paradigmatic candidates for such entities. But in contrast to the widely shared view that there are objects that exist, contemporary philosophers have considerably diverging and competing views about what objects or what kinds of objects exist. Indeed, many debates in metaphysics or ontology turn around the question of what there is, i.e., what kinds or types of objects exist.[1] A classical example is the debate about the question whether only concrete particulars exist or also abstract universals such as properties.[2] Another example is the contemporary debate about the question of what kinds of composite objects exist.[3]

The topic of this book is not the question of what there is or exists. My aim is not to defend a particular view on what objects or kinds of objects exist. Rather, I am interested in the more fundamental question of what it is, for an object that exists, to exist, i.e., what the existence of an object consists in.[4] Thus, my aim is to answer the question as to what it is for an object that exists to exist.

This question arises for any view which maintains that there are objects which exist. It is more fundamental than the question of what there is or exists because any answer to the latter question presupposes an answer to the question

---

1 I use the terms 'metaphysics' and 'ontology' interchangeably unless otherwise noted.

2 Good overviews for different periods of time can be found in Gabriele Galluzzo and Riccardo Chiaradonna, *Universals in Ancient Philosophy* (Pisa: Edizioni della Normale, 2013); Gabriele Galluzzo and Michael J. Loux, *The Problem of Universals in Contemporary Philosophy* (Cambridge: Cambridge University Press, 2015); Paul Vincent Spade, *Five Texts on the Mediaeval Problem of Universals: Porphyry, Boethius, Abelard, Duns Scotus, Ockham* (Indianapolis; Cambridge: Hackett, 1994).

3 Cf. Johannes Hübner, *Komplexe Substanzen* (Berlin; New York: Walter de Gruyter, 2007); Daniel Z. Korman, *Objects: Nothing Out of the Ordinary* (Oxford: Oxford University Press, 2015); Giorgio Lando, *Mereology: A Philosophical Introduction* (London; New York: Bloomsbury Academic, 2017).

4 The need to address this question has recently been acknowledged by analytic philosophers such as Peter van Inwagen who—following the tradition of W.V.O. Quine—used to be more focused on the question of what there is or exists, cf. Peter van Inwagen, "Being, Existence, and Ontological Commitment," in *Metametaphysics: New Essays on the Foundations of Ontology*, ed. David J. Chalmers, David Manley, and Ryan Wasserman (Oxford: Clarendon Press, 2009). Van Inwagen is currently working on a book on being which intends to expand on this issue. He was kind enough to share with me a draft of this book to let me see in which direction his thought is developing, and I am very grateful to him.

https://doi.org/10.1515/9783110979879-003

of what it is to exist. As soon as someone entertains a view on what there is or exists, they are confronted with the problem of explaining what it is for that which exists to exist. In other words, any view which maintains that there are objects which exist presupposes a view on what it is for an object that exists to exist. For example, a view which maintains that objects like dogs, cats, trees, puddles of water, chunks of gold, and human beings exist has to explain what it is for dogs, cats, trees etc. to exist, i.e., what the existence of such objects consists in. And a view which maintains that only elementary particles exist has to explain what it is for elementary particles to exist, i.e., what the existence of elementary particles consists in.

Again, as soon as one affirms *that* there are objects which exist, the problem of explaining what it is for such objects to exist arises, regardless of *what* the objects are that exist. So, the view on what it is for that which exists to exist which I will defend in this book is, to a certain extent, *neutral* with respect to what it is that is supposed to exist. Nevertheless, it is difficult to inquire into the question of what it is for that which exists to exist in abstraction from the question of what it is that is supposed to exist. Thus, for the sake of illustration, I will use examples of objects which I believe have a good chance to be widely accepted as objects that exist, but nothing in my argument hinges on any of the chosen examples.

What the topic of this book is and what it is not can also be explained by making use of a terminological distinction which is currently in vogue, namely, the distinction between ontology or metaphysics and metaontology or metametaphysics.[5] Put roughly, 'ontology' is the study that attempts to answer the question 'What exists or is there, i.e., what objects or kinds of objects exist?' and 'metaontology' is the study that attempts to answer the question 'What is it for that which exists to exist, i.e., what does the existence of that which exists consists in?'.[6] In other words, ontology is concerned with beings (*entia*) while metaontol-

---

5 I use the terms 'metaontology' and 'metametaphysics' interchangeably unless otherwise noted.

6 This is just one way to make sense of the distinction. The question of how best to distinguish between ontology or metaphysics and metaontology or metametaphysics is itself a matter of controversy, see Francesco Berto and Matteo Plebani, *Ontology and Metaontology: A Contemporary Guide* (London: Bloomsbury Academic, 2015); Ricki Bliss and J. T. M. Miller, eds., *The Routledge Handbook of Metametaphysics* (London: Routledge, 2020); David J. Chalmers, David Manley, and Ryan Wasserman, eds., *Metametaphysics: New Essays on the Foundations of Ontology* (Oxford: Clarendon Press, 2009); Tuomas E. Tahko, *An Introduction to Metametaphysics* (Cambridge: Cambridge University Press, 2015). My way to explain the distinction between ontology and metaontology follows roughly van Inwagen, "Being, Existence, and Ontological Commitment," 472–475.

ogy is concerned with the being of beings, i.e., with what it is for beings to be (*esse*).

Again, metaontology is more fundamental than ontology because the former investigates into something that the latter takes for granted, namely, that a being is a '*be*-ing' (*ens*). And metaontology is *neutral* with respect to ontology precisely because it considers beings as beings, i.e., in abstraction from what kind or type of being they are. It does not matter whether that which exists is a tree, a human being, an elementary particle, or anything whatsoever because it is not considered as *what* it is but as something *that* is or exists. Thus, the topic of my book can be described as metaontological and not ontological in nature. According to this terminology, I defend a particular metaontological view on what it is for that which exists to exist against several rival metaontological views.

In general, I have no quarrel with this way to delineate my project. But a disadvantage of this approach is that there is no agreement on what metaontology is and how it is best distinguished from ontology. So, there is danger that the distinction may confuse matters more than that it helps to clarify them. For example, Markus Gabriel characterizes 'ontology' as the study which inquires into the question of what existence or being is.[7] But this is precisely what van Inwagen thinks the job of 'metaontology' is.[8]

Furthermore, the impression of a need for 'metaontology' seems to have arisen from a narrow characterization of what 'ontology' is concerned with. Now, one who never embraced such a narrow view on ontology or metaphysics might wonder why they should accept that a whole new discipline like 'metaontology' or 'metametaphysics' is needed. For example, for philosophers who take inspiration in the works of a medieval author like Thomas Aquinas, metaphysics or ontology is not so much concerned with the question of what there is or exists but rather investigates into the being or existence of that which exists.[9] Thus,

7 Cf. Markus Gabriel, *Fields of Sense: A New Realist Ontology* (Edinburgh: Edinburgh University Press, 2015), 37; *Fiktionen* (Berlin: Suhrkamp Verlag, 2020), 24; *Sinn und Existenz: Eine realistische Ontologie* (Berlin: Suhrkamp, 2016), 25–26.

8 Cf. van Inwagen, "Being, Existence, and Ontological Commitment," 475.

9 Cf. *In Meta*, prooemium: "Ex quo apparet, quod quamvis ista scientia praedicta tria consideret, non tamen considerat quodlibet eorum ut subiectum, sed ipsum solum ens commune. Hoc enim est subiectum in scientia, cuius causas et passiones quaerimus, non autem ipsae causae alicuius generis quaesiti. Nam cognitio causarum alicuius generis, est finis ad quem consideratio scientiae pertingit. Quamvis autem subiectum huius scientiae sit ens commune, dicitur tamen tota de his quae sunt separata a materia secundum esse et rationem. Quia secundum esse et rationem separari dicuntur, non solum illa quae nunquam in materia esse possunt, sicut Deus et intellectuales substantiae, sed etiam illa quae possunt sine materia esse, sicut ens commune. Hoc tamen non contingeret, si a materia secundum esse dependerent. Secundum

what van Inwagen, for example, considers to be a central metaontological question is for those philosophers at the core of a metaphysical inquiry.

It is important to note that with these remarks I do not question the possibility or value of metaontology or metmetaphysics in general. I just want to hint at the difficulties and problems which might arise by using this terminology to justify a different way to explain what this book is about and what not. This can be done without introducing a distinction between ontology or metaphysics and metaontology or metametaphysics. My hope is that this alternative way stays clear of controversial terminological and methodological issues which might be more confusing than helpful.

According to this alternative approach, there are three different questions which can be asked in connection with the existence of objects.[10] The first question asks, 'What is there, i.e., what objects or kinds of objects exist?'. This question can be called the 'inventory question'.[11] My book is not concerned with this question. I simply assume that there are objects which exist, and my aim is not to contribute in any essential respect to the debate about what objects or what kinds of objects exist.

A second question asks, 'What is it for that which exists to exist, i.e., what does the existence of that which exists consists in?'. Let us call this question—for lack of a better name—the 'the nature question'.[12] The nature question inquires not into what objects or kinds of objects exist but into the existence of that which exists. In other words, the nature question is not interested in determining what there is or exists but is interested in determining what it is, for that which exists, to exist.

igitur tria praedicta, ex quibus perfectio huius scientiae attenditur, sortitur tria nomina. Dicitur enim scientia divina sive theologia, inquantum praedictas substantias considerat. Metaphysica, inquantum considerat ens et ea quae consequuntur ipsum. Haec enim transphysica inveniuntur in via resolutionis, sicut magis communia post minus communia. Dicitur autem prima philosophia, inquantum primas rerum causas considerat." For a more detailed exposition of Aquinas's view on the subject and nature of metaphysics, see Gaven Kerr, "The Meaning of *Ens Commune* in the Thought of Thomas Aquinas," *Yearbook of the Irish Philosophical Society* (2008); John F. Wippel, *The Metaphysical Thought of Thomas Aquinas: From Finite Being to Uncreated Being* (Washington, D.C.: Catholic University of America Press, 2000), 3–22.

10 The idea for this approach is taken over from William F. Vallicella but I formulate the questions in a different way, cf. William F. Vallicella, "Existence: Two Dogmas of Analysis," in *Neo-Aristotelian Perspectives in Metaphysics*, ed. Daniel D. Novotny and Lukas Novak (New York; London: Routledge, 2014), 45–46; *A Paradigm Theory of Existence: Onto-Theology Vindicated* (Dordrecht: Kluwer Academic, 2002), xi, 27–28.

11 Cf. "Existence: Two Dogmas of Analysis," 45–46.

12 Cf. *A Paradigm Theory of Existence*, xi, 1, 27. I use the terms 'existence' and 'being' interchangeably unless otherwise noted.

Strictly speaking, the nature question does not ask what existence or being is, i.e., what the nature of existence or being is.[13] From the point of view which I will expound and defend later in more detail, any attempt to offer a kind of conceptual analysis of the notion of existence (*esse*) is doomed to fail because existence is something primitive which cannot be analyzed any further.[14] The essence or nature of existence is existence. Existence is simply that by which that which exists exists.[15]

But this does not imply that the existence *of an object*, i.e., an object's *having* existence is primitive. It is one thing to ask what the nature of existence is, but it is another thing to ask what the nature of the relation is between an object and the existence it has.[16] So, it is important to have in mind that the nature question does not ask 'What is existence?' but 'What is it for that which exists to exist, i.e., what does the existence of an object consist in?'. What the nature question looks for is an explanation of the existence *of an object* that exists. For example, according to a view which I will deal with later in more detail, the existence of an object consists in that object's having the property of existing. So, according to this view, existence is a property of an object that exists. In this case, what explains the existence of an object is that the property of existing is among the properties it has. An object that exists exists in virtue of having the property of existing.

A third and final question which one can ask about the existence of objects is 'Why does that which exists exist?'. Let us call this question the 'ground question'.[17] We must be careful here, because the 'why' is ambiguous. To see this ambiguity better, let us return to the just-mentioned view that the existence of an object consists in that object's having the property of existing. As we have just seen, this view gives an answer to the nature question. Now, in a certain sense, a view which answers the nature question also answers the ground question: Why does that which exists exist? Well, according to the view under consideration, the existence of an object consists in that object's having the property of

---

13 Here I part company with authors like Vallicella or van Inwagen, who state that the nature question asks what existence or being is, cf. ibid., 1; van Inwagen, "Being, Existence, and Ontological Commitment," 472–475.

14 For a more detailed explanation of why conceptual analysis does not work in the case of existence or *esse*, see Gaven Kerr, "Thomistic *Esse* and Analytical Philosophy," *International Philosophical Quarterly* 55, no. 1 (2015).

15 Cf. Vallicella, *A Paradigm Theory of Existence*, 5.

16 Cf. Gaven Kerr, *Aquinas's Way to God: The Proof in De Ente et Essentia* (Oxford: Oxford University Press, 2015), 102.

17 Cf. Vallicella, *A Paradigm Theory of Existence*, xi, 27.

existing. An object exists *in virtue of* having the property of existing. So, in a sense, an answer to the nature question explains why an object exists. Let us call the ground question which is satisfactorily answered by an answer given to the nature question the 'weak ground question'. The weak ground question is just another way to formulate the nature question. Thus, if the question 'What is it for that which exists to exist?' is answered, the weak version of the question 'Why does that which exists exist?' is answered as well.

However, some might object that the ground question requires a different kind of explanation. Even if the nature question is answered, the objector would say, there remains something to be explained about the existence of an object which is not explained by a view on what it is for that which exists to exist. For example, in the case of the view that the existence of an object consists in that object's having the property of existing, the view does not explain in virtue of what the object has the property of existing. The view just explains that an object exists in virtue of having the property of existing, but it does not explain in virtue of what the object has that property.

Look at it this way: If an object has a property that it might as well not have, an explanation is needed for the fact that it has that property. Thus, unless one wants to argue that existence is an essential property of that which exists and that existing objects therefore exist necessarily, an explanation is required for the contingent fact that an object that exists has the property of existing so that it exists.

Understood in this way, the ground question asks why that which has existence has existence although it might as well lack existence. The 'why' asks for a kind of causal explanation for the existence of that which exists only contingently. Such an explanation is not given by an answer to the nature question. Let us call the ground question which is not satisfactorily answered by an answer to the nature question the 'strong ground question'. One who asks this question is looking for an answer to the so-called 'ultimate why question': 'Why is there anything at all rather than nothing whatsoever?'.[18]

In this book, I am concerned with the nature or weak ground question and not with the strong ground question. I am not trying to explain why there is something rather than nothing, i.e., I do not look for a kind of causal explana-

18 For two recent collections of essays which revolve around this question, see Tyron Goldschmidt, *The Puzzle of Existence: Why Is There Something Rather Than Nothing?* (New York: Routledge, 2013); John F. Wippel, *The Ultimate Why Question: Why Is There Anything At All Rather Than Nothing Whatsoever?* (Washington, D.C.: Catholic University of America Press, 2011).

tion for the existence of that which exists only contingently.[19] Rather, given *that* there are contingently existing objects, I am interested in the question of what it is for such objects to exist. What does their existence consist in? For example, does the existence of an object X consist in X's having the property of existing? Or does the existence of an object X consist in something else's having a property, e.g., a property's having the property of being instantiated or a domain's having the property of containing X and being [X]ish? To sum up: My inquiry is about the problem of explaining what it is for that which exists to exist, which arises for any view that maintains *that* there are objects which exist.

In particular, I defend two theses by which I hope to advance the contemporary debate about the nature question. First, I argue that what many contemporary philosophers have to say about what it is for an existing object to exist is unsatisfactory. Their views about what the existence of an object consists in are riddled with problems. In other words, there are good reasons to reject the answers they give to the nature question.

Second, I claim that Thomas Aquinas answered the nature question in a better way. His view explains in a more satisfactory way than contemporary perspectives what it is for an object to exist, i.e., what the existence of an object consists in. Furthermore, the consideration of Aquinas's view contributes a new perspective and vantage point which allows one to see why the contemporary debate about the nature question has reached an impasse.

My book is structured into two parts to substantiate these two claims. The first part consists of five chapters. It offers a systematic reconstruction of the contemporary debate about the nature question. I reconstruct five views on what it is for an object to exist which have been entertained in the contemporary discussion. I show that each of them is unable to explain in a satisfactory way what it is for an object to exist. The second part contains two chapters. Chapter 6 reconstructs Aquinas's explanation of what it is for an object to exist. The last chapter explains why his view can solve the problems which the views discussed in the first part cannot.

Let me offer a more detailed overview of the seven chapters which are about to follow. Chapter 1 deals with the view I have briefly mentioned above to illustrate the difference between the inventory, the nature, and the ground question. According to this view, for an object to exist is to have the property of existing. In other words, the existence of an object consists in that object's having the prop-

---

19 For two recent excellent studies on how Aquinas answers the strong ground question, see Gaven Kerr, *Aquinas and the Metaphysics of Creation* (New York: Oxford University Press, 2019); *Aquinas's Way to God.*

erty of existing. Thus, the existence of an object is explained by that object's having the property of existing.

What distinguishes this view from the other four views I deal with in the first part is that it employs a *non-reductive* explanatory strategy. This explanatory strategy is non-reductive in nature because this view considers existence to be a property *sui generis* of an object that has it. For the sake of convenience, let us call this view the 'non-reductive-object view'. It has been embraced in different variants by authors such as Francesco Berto, Saul A. Kripke, Colin McGinn, Barry Miller, Terence Parsons, Graham Priest, Richard Routley, Timothy Williamson, Palle Yourgrau, and Edward N. Zalta.[20]

I argue that such a view is unsatisfactory for three reasons. First, its answer to the nature question is viciously circular because an object can have or possess the property of existing only if it exists. The idea that for an object to exist is to have the property of existing does not explain, but rather presupposes, the existence of that object.

Second, if existence were a property of an object that exists, that existence would be either an essential or an accidental property of that object, i.e., it would be either a property which was necessary to the object, or a property which that object had but which was not necessary to that object.[21] But existence is neither necessary nor unnecessary for the object.

---

20 Cf. Francesco Berto, *Existence as a Real Property: The Ontology of Meinongianism*, (Dordrecht; London: Springer, 2013); Saul A. Kripke, *Reference and Existence: The John Locke Lectures* (New York: Oxford University Press, 2013); Colin McGinn, *Logical Properties: Identity, Existence, Predication, Necessity, Truth* (Oxford: Clarendon Press, 2000); Barry Miller, *The Fullness of Being: A New Paradigm for Existence* (Notre Dame, Indiana: University of Notre Dame Press, 2002); Terence Parsons, *Nonexistent Objects* (New Haven; London: Yale University Press, 1980); Graham Priest, *Towards Non-Being: The Logic and Metaphysics of Intentionality*, 2 ed. (Oxford: Oxford University Press, 2016); Richard Sylvan, *Exploring Meinong's Jungle and Beyond: An Investigation of Noneism and the Theory of Items* (Canberra: Australian National University, 1980); Timothy Williamson, "Necessary Existents," *Royal Institute of Philosophy Supplements* 51 (2002); Palle Yourgrau, *Death and Nonexistence* (New York: Oxford University Press, 2019); Edward N. Zalta, *Abstract Objects: An Introduction to Axiomatic Metaphysics* (Dordrecht; Lancaster: D. Reidel, 1983). For helpful explanations on what differentiates the views of these authors, see Kerr, *Aquinas's Way to God*; "Thomistic *Esse* and Analytical Philosophy."; Andreas Luckner and Sebastian Ostritsch, *Existenz* (Berlin; Boston: Walter de Gruyter, 2018); Peter van Inwagen, "McGinn on Existence," *The Philosophical Quarterly* 58, no. 230 (2008).

21 I am thankful to Eleonore Stump, who made me aware that I need to briefly explain how I characterize essential and accidental properties. I clarify below in more detail how essential and accidental properties are standardly characterized in modal terms in contemporary analytic philosophy (see 1.2.2).

Existence is not an essential property of that which exists, because this would mean that everything that exists exists necessarily and consequently could not possibly not exist, which is obviously false.[22] No object which has existence exists in every possible world.[23] Nor is existence an accidental property of an object that exists, because this would mean that existence is a property which that object can lose without ceasing to exist. This is obviously false because an object that exists cannot lose existence and continue to exist.

Finally, the view that the existence of an object consists in that object's having a property *sui generis*—the property of existing—does not explain what is supposed to be explained, namely, what it is for an object to exist. It is intelligible what it is for an object to have an accidental or an essential property. But it is unintelligible what it is for an object to have a property *sui generis* which is neither essentially nor accidentally related to an object which has it. Thus, the attempt to explain what it is for an object to exist by making use of the idea of existence as a property *sui generis* of an object that exists is unsatisfactory, because it does not explain how that which exists is related to existence as its property. In other words, it is not explained how existence as a property *sui generis* determines the object which has it with respect to its existence if existence as a property *sui generis* determines its object neither essentially nor accidentally as any other property which determines an object that has it.

The remaining four chapters of the first part of my book reconstruct and criticize four views which employ a *reductive* explanatory strategy. This explanatory strategy is reductive in nature because the related views try to identify and reduce the existence of an object with and to something having a property other

---

22 It should be noted that a philosopher like Timothy Williamson might be willing to accept that existence is an essential property of an object that exists because he defends the highly controversial hypothesis that everything that exists exists necessarily, i. e., that there is a plurality of necessary existents, cf. Williamson, "Necessary Existents." I do not engage with this minority view because authors like Theodore Sider, Joshua Spencer, or Ian Rumfitt have convincingly shown that Williamson's argument is flawed, see Theodore Sider, "Williamson's Many Necessary Existents," *Analysis* 69, no. 2 (2009); Joshua Spencer, "Unnecessary Existents," *Canadian Journal of Philosophy* 43, no. 5/6 (2013); Ian Rumfitt, "Contingent Existents,"*Philosophy* 78, no. 4 (2003).
23 One should not infer from this remark that I am committed to a possible world semantics or metaphysics. I just use the familiar and established idea of possible worlds as a heuristic device. Furthermore, it is important to note that the claim that no object which has existence exists necessarily does not rule out the possibility that there is a being— usually called 'God'—which exists necessarily. If there is such a being, it exists necessarily because it does not *have* existence but *is* existence itself, i. e., it is a being whose essence is identical with its existence. In other words, if there is such a being, it does not exist in virtue of something it *has*—existence—but in virtue of what it *is* essentially, namely, existence.

than existence. What distinguishes these views from each other is that with which they attempt to identify the object's existence and to which they attempt to reduce the object's existence.

In Chapter 2, I begin with a view according to which for an object to exist is to have a property. But the relevant property is not existence. Rather, it is an object's property of being numerically identical with itself. Existence is not a property *sui generis* of an object but is identical with and can be reduced to another property of an object, namely, an object's property of being identical with itself. Existence is self-identity. In other words, the existence of an object consists in that object's property of being numerically identical with itself. Thus, what explains the existence of an object is that this object has the property of being identical with itself. In the contemporary debate, such a view has been embraced by Peter van Inwagen, who will be my main interlocutor in this chapter.[24] For the sake of convenience, let as call this view the 'reductive-object view', or alternatively 'van Inwagen's view', without losing sight of the fact that this approach has also been embraced by others.[25]

I level three objections against such a view. First, it fails to show that the existence of an object can be reduced to an object having the property of being identical with itself by its own standards for a successful reduction. Second, if for an object to exist is to have the property of being identical with itself, it follows that every object which has the property of being identical with itself exists necessarily because self-identity is an essential property of objects. But objects which have existence and therefore exist do not exist necessarily. They could possibly not have existence and not exist. Finally, the view is vulnerable to counterexamples. Not all objects which are identical with themselves exist. For example, fictional objects like Sherlock Holmes have the property of being identical with themselves but do not exist.[26] Such counterexamples indicate that the exis-

---

24 Cf. van Inwagen, "Being, Existence, and Ontological Commitment."; *Metaphysics*, 4 ed. (Boulder, CO: Westview Press, 2014), 291–326.

25 According to Reinhardt Grossmann, Gottlob Frege defended such a view in his *Dialog mit Pünjer über Existenz*, cf. Reinhardt Grossmann, *The Existence of the World: An Introduction to Ontology* (London: Routledge, 1992), 99–101. According to Andreas Luckner and Sebastian Ostritsch, it is important to distinguish Frege's view that existence is an object's property of being numerically identical with itself from Saul A. Kripke's view that existence is an object's property of being identical with something that there is or exists, cf. Luckner and Ostritsch, *Existenz*, 41–42. In this book, I deal with the former and not with the latter view.

26 To be more precise, fictional entities like Sherlock Holmes do not exist in the sense of being actual. In other words, Sherlock Holmes exists but exists only in a qualified sense. But as I will explain below in more detail, in this inquiry I am only interested in existence in the sense of actuality. I am thankful to Jeremy Skrzypek for drawing my attention to this point.

tence of an object does not consist in that object having the property of being identical with itself.

The remaining three views I discuss in the first part also employ a reductive explanatory strategy. What distinguishes them from the reductive-object view is that they reject the idea that the existence of an object consists in an object's property of being identical with itself. The existence of an object cannot be identified with and reduced to an object having the property of being numerically identical with itself or any other property which that object has. Rather, what unites these three views is the idea that the existence of an object is identical with and can be reduced to a property which *something else* has. In other words, these views maintain that the existence of an object consists in a property of something else. To be more precise, they claim that the existence of an object consists in something else having a property.

Chapter 3 discusses the view that the existence of an object consists in a property's property of being instantiated. Thus, what explains the existence of an object is that a property has the property of being instantiated. In the contemporary debate, this view is often considered to be the dominant or orthodox view within analytic philosophy.[27] Furthermore, it is standardly associated with a tradition of thinkers which includes Immanuel Kant, Gottlob Frege, Bertrand Russell, and W.V.O. Quine.[28] For the sake of convenience, let us call this view the 'property view'.[29] Now, it is important to note that in this book I am not interested in the exegetical question whether any of these philosophers in fact entertained such a view or in which way the different authors held different views. Rather, I deal with the view insofar as it is a systematic option in the debate

27 For example, see Christian Kanzian, "Sein als Existenz," in *Handbuch Ontologie*, ed. Jan Urbich and Jörg Zimmer (Berlin: Springer-Verlag, 2020), 351; Gyula Klima, "Aquinas' Reception in Contemporary Metaphysics," in *The Oxford Handbook of the Reception of Aquinas*, ed. Matthew Levering and Marcus Plested (Oxford: Oxford University Press, 2021), 566; McGinn, *Logical Properties*, 17; Turner C. Nevitt, "How to Be an Analytic Existential Thomist," *The Thomist* 82, no. 3 (2018): 322, 324.

28 Cf. Klima, "Aquinas' Reception in Contemporary Metaphysics," 566; Kris McDaniel, *The Fragmentation of Being* (Oxford: Oxford University Press, 2017), 55–56; van Inwagen, *Metaphysics*, 296–300.

29 It could be objected that I am lumping together different views. For example, Russell did not argue that existence is a property of properties, but that existence is a property of propositional functions. But even authors like van Inwagen, who want to distinguish between these views, admit that they are very similar, cf. "McGinn on Existence," 38. Because nothing in what follows rests on whatever differences there might be between the views entertained by Frege, Russell, or Quine, I lump them together under the umbrella view that existence is a property of properties. I leave out Kant because it is a matter of controversy in which sense and to what extent Kant anticipated what Frege, Russell, or Quine would have to say about existence.

and a possible answer to the nature question which is discussed in this form by contemporary philosophers.[30]

The property view is unsatisfactory because it generates four problems. To begin with, the explanation of this view for what the existence of an object consists in is viciously circular, because a property's having the property of being instantiated presupposes that only existing objects are instances of the relevant property. In other words, the existence of an object must be presupposed to make sense of the idea that a property has the property of being instantiated. A property's having the property of being instantiated requires that there is at least one instance of this property. But the idea that there is an instance of a property implies that this instance is an instance of the property because it exists. Something which does not exist cannot be an instance of a property. Thus, instantiation does not explain but rather presupposes existence.

Second, the property view triggers a vicious infinite regress as soon as it tries to explain what it is for the property to exist whose having the property of being instantiated is supposed to explain what it is for a particular object to exist. To explain what it is for a property A to exist, the view must introduce another property B which has the property of being instantiated by A, i.e., of having A as an instance. Thus, the existence of A consists in B having the property of being instantiated by A, i.e., of having A as an instance. But now everything repeats itself, because to explain what it is for B to exist, the view must introduce a further property C which has the property of being instantiated by B, i.e., of having B as an instance and so on *ad infinitum.*

Third, this view faces the problem that it works only within a heavily inflated ontology. If the existence of an object consists in a property having the property of being instantiated, the view needs to introduce an infinite or at least very large number of properties which themselves exist to be able to explain the existence of the objects that exist.

Finally, the view that the existence of an object consists in a property's having the property of being instantiated is confronted with the problem of the co-existence of objects and properties. The property view cannot maintain that the properties which are supposed to account for the existence of the objects that exist exist alongside these objects. Nor can it claim that these properties inhere in the objects that exist. Consequently, the view is forced to state that objects and the properties which are supposed to account for the existence of the objects exist in two different ontological realms. The problem of co-existence might be

---

30 Cf. Kerr, *Aquinas's Way to God*, 68–76; McGinn, *Logical Properties*, 17–37; Vallicella, *A Paradigm Theory of Existence*, 93–125; van Inwagen, *Metaphysics*, 296–300.

solved in this way. However, its solution comes at the price of a dualistic ontology, which makes the view unattractive.

The last two views I address in the first part of my book agree with the property view that the existence of an object consists in something else's having a property. They part company with the property view when it comes to the question of what kind of thing or entity it is whose having a property is supposed to explain what it is for an object to exist. What unites the two views I deal with in Chapters 4 and 5 is the idea that the existence of an object X does not consist in the property of a *property* of being instantiated by X but in the property of a *domain of objects* of containing X and being [X]ish. Thus, according to these views, what explains the existence of an object X is that a domain has the property of containing X and being [X]ish. Now, while the two views agree on what kind of thing or entity it is whose having a property is supposed to explain the existence of an object—a domain of objects—they disagree about whether this domain of objects is unrestricted or restricted. For the sake of brevity, let us call an unrestricted domain of objects a 'UDO' and a restricted domain of objects a 'REDO'.

Chapter 4 starts with the view that the existence of an object X consists in the property of a UDO of containing X and being [X]ish. Thus, what explains the existence of an object is that a UDO has the property of containing X and being [X]ish. A UDO is considered to be unrestricted in the sense that it is a domain of objects which contains everything that exists. Whatever a UDO is identified with—e.g., mind-independent reality, being, or the world—it is characterized as a *totality* which contains everything which exists. In the contemporary debate about the nature question, it has been argued that philosophers such as Kant, Milton Munitz, Bruce Aune, Nicolai Hartmann, and Frederic Sommers subscribe to versions of such a view.[31] For the sake of convenience, let us call this view the 'unrestricted-domain view'. Again, it is important to have in mind that in this book I am not concerned with the exegetical question of whether it is true that any of the mentioned philosophers really embraced such a view. Rather, I am addressing this view because it represents a systematic option in the debate and provides a possible answer to the nature question which contemporary philosophers discuss.[32]

The view that the existence of object X consists in a UDO having the property of containing X and being [X]ish is unsatisfactory for two reasons. First, the view's explanation of what it is for an object to exist is viciously circular because

31 Cf. Gabriel, *Sinn und Existenz*, 98–123, 138–139; Vallicella, *A Paradigm Theory of Existence*, 127.
32 Cf. Gabriel, *Sinn und Existenz*, 141–157, 224–270; Vallicella, *A Paradigm Theory of Existence*, 127–157.

a UDO's having the property of containing X and being [X]ish presupposes that X exists. The existence of X must be presupposed to make sense of the idea that a UDO has the property of containing X. Something which does not exist cannot be contained in a UDO. Thus, containment does not explain but rather presupposes the existence of that which is contained.

Second, the unrestricted-domain view has the problem that it cannot even be formulated in a coherent way. This approach must assume that a UDO exists because it claims that the existence of an object X consists in a UDO having the property of containing X and being [X]ish. Something which does not exist cannot have the property of being [X]ish. However, this view must also negate that a UDO exists, because the existence of a UDO would consist in a domain of objects D's property of containing the UDO and being [UDO]ish. To be more precise, the existence of a UDO would consist in D's having the property of containing the UDO and being [UDO]ish. Such a domain of objects D must exist because a domain of objects which does not exist cannot have the property of containing a UDO and being [UDO]ish. But the existence of such a domain D is ruled out by the characterization of a UDO as a domain of objects which contains everything that exists. Thus, the explanation of what it is for an object X to exist in this view requires the existence of a UDO whose existence this view must deny because it is impossible that there exists a domain of objects D which contains a UDO.

In Chapter 5, I discuss the view that the existence of an object X consists in the property of a restricted domain of objects—a REDO—of containing X and being [X]ish. Thus, according to this view, what explains the existence of an object X is that a REDO has the property of containing X and being [X]ish. A REDO is considered to be restricted in the sense that it is a domain of objects which contains only some of the objects—including X—that exist. Consequently, there must be other restricted domains of objects to account for the existence of the objects that the REDO under consideration does not contain. My main interlocutor in this chapter will be Markus Gabriel, who has presented and defended such a view in recent years and whose approach has been widely discussed even outside the German speaking philosophical community.[33] For the sake of

33 See, for example, Jocelyn Benoist, *L'Adresse du Réel* (Paris: Librairie Philosophique J. Vrin, 2017); Thomas Buchheim and Markus Gabriel, eds., *Neutraler Realismus* (Freiburg im Breisgau: Karl Alber, 2016); Markus Gabriel, ed. *Der Neue Realismus* (Berlin: Suhrkamp, 2014); "Existenz, realistisch gedacht," in *Der Neue Realismus*, ed. Markus Gabriel (Berlin: Suhrkamp, 2014); *Fields of Sense*; *Fiktionen*; "The Meaning of ‚Existence' and the Contingency of Sense," *Frontiers of Philosophy in China* 9, no. 1 (2014); "Metaphysik oder Ontologie?,"*Perspektiven der Philosophie: Neues Jahrbuch* 42 (2016); "Nachwort: Abgesang und Auftakt," in *Angst vor der Wahrheit: Ein Plä-*

convenience, let us call this view the 'restricted-domain view', or alternatively 'Gabriel's view' without losing sight of the fact that similar views have been advanced, for example, by authors such as Jason Turner and Kris McDaniel.[34]

The position that the existence of object X consists in a REDO having the property of containing X and being [X]ish is unsatisfactory for six reasons. First of all, the view triggers a vicious infinite regress as soon as it tries to explain what it is for the REDO to exist whose having the property of containing X and being [X]ish is supposed to explain what it is for X to exist. To explain what it is for a $REDO_1$ to exist, the view must introduce another $REDO_2$ which has the property of containing the $REDO_1$ and being [$REDO_1$]ish. But now everything repeats itself because to explain what it is for the $REDO_2$ to exist, the view must introduce a further restricted domain of objects $REDO_3$, which has the property of containing the $REDO_2$ and being [$REDO_2$]ish, and so on *ad infinitum*.

Second, the restricted-domain view faces the problem that it works only within a heavily inflated ontology. If the existence of an object X consists in a REDO having the property of containing X and being [X]ish, the approach must introduce an infinite number of restricted domain of objects which themselves exist to explain the existence of only one object that exists.

Third, the view's explanation of what it is for an object X to exist is viciously circular, because a REDO's having the property of containing X and being [X]ish presupposes that X exists. The existence of X must be presupposed to make sense of the idea that a REDO has the property of containing X. Something which does not exist cannot be contained by a REDO. Thus, containment does not explain but rather presupposes the existence of that which is contained.

Fourth, the restricted-domain view has the problem that a REDO's having the property of containing X and being [X]ish is not constitutive of the existence of X. For example, it is not constitutive of the existence of Joe Biden that the restricted domain of objects which is the White House has the property of containing Biden and being [Joe Biden]ish. Even worse, if there were a REDO whose having the

doyer gegen Relativismus und Konstruktivismus, Paul A. Boghossian (Berlin: Suhrkamp, 2013); *Neo-Existentialism: How to Conceive of the Human Mind After Naturalism's Failure* (Cambridge: Polity, 2018); "Neutral Realism," *The Monist* 98, no. 98 (2015); *Sinn und Existenz*; "Sinnfeldontologie oder reformierte Metaphysik? Replik auf Le Moli,"*Perspektiven der Philosophie: Neues Jahrbuch* 42 (2016); *Warum es die Welt nicht gibt* (Berlin: Ullstein, 2013); *Why the World Does Not Exist* (Cambridge: Polity Press, 2015); Peter Gaitsch, Sandra Lehmann, and Philipp Schmidt, eds., *Eine Diskussion mit Markus Gabriel: Phänomenologische Positionen zum Neuen Realismus* (Wien: Verlag Turia + Kant, 2017).

34 Cf. McDaniel, *The Fragmentation of Being*; Jason Turner, "Ontological Pluralism," *The Journal of Philosophy* 107, no. 1 (2010).

property of containing Joe Biden is constitutive of his existence, it would result that Biden does not exist as that which we believe him to exist as, namely, as an object which exists in its own right. If Joe Biden's existence consists in a REDO's having the property of containing him and being [Joe Biden]ish, Biden does not exist as an object which exists in its own right. Rather, he exists only as an object which exists in the REDO as a part of it, or he inheres in the REDO as an accidental form of it. Thus, the restricted-domain view is not only unable to explain what the existence of an object which exists in its own right consists in. It must also negate that it is possible for such objects to exist. Objects like Joe Biden lose their ontological status as objects which exist in their own right and become objects which exist only in restricted domains of objects as parts or accidental properties of them.

The fifth problem with this view is that its attempt to identify and reduce the existence of an object X with and to the REDO's property of containing X and being [X]ish does not work.[35] The reductionism does not work because every object that exists exists as an object which has accidental properties and the restricted-domain view cannot accommodate this fact.[36] For example, this view cannot explain what it is for *sitting* Joe Biden to exist. To explain what it is for *sitting Joe Biden* to exist, the view must argue that the existence of sitting Joe Biden consists in a REDO's having the property of containing sitting Joe Biden and being [sitting Joe Biden]ish. For the sake of illustration, let the relevant REDO be the White House. In this case, this view's explanation of what it is for sitting Joe Biden to exist is that the White House has the property of containing sitting Joe Biden and being [sitting Joe Biden]ish.

But the problem with this explanation is that the existence of sitting Biden cannot be identified with and reduced to a property which something else has, e.g., the White House. The reduction does not work because Joe Biden's having the property of sitting cannot be identified with and reduced to a property which the White House has. It would be absurd to claim that the White House has the

35 The idea for this objection is taken from Vallicella, but I do not follow exactly his way of formulating the objection, see Vallicella, *A Paradigm Theory of Existence*, 132–135. The objection also applies to the view that the existence of an object X consists in a UDO's having the property of containing X and being [X]ish.

36 Again, it is important to note that if God exists as an entity whose essence is identical with its existence, he is an exception because he is an object that exists, but which does not have any accidental properties. An entity whose essence is identical with its existence is metaphysically simple, i.e., it is impossible that it has accidental properties because having accidental properties requires metaphysical composition. For a more detailed argument, see, for example, *ST*, I, q. 3, a. 6. I bracket the issue of the possible existence of God so as to not make my argument more complicated than necessary.

property of sitting. Rather, if the White House has the property of containing *sitting* Joe Biden, it is in virtue of Joe Biden's having the property of sitting. And Joe Biden can have the property of sitting only if he exists. Thus, the existence of an object X that has accidental properties (e. g., sitting Joe Biden) cannot be identified with and reduced to a REDO's having the property of containing X and being [X]ish, because having these accidental properties cannot be identified with and reduced to properties which the REDO has.

A sixth and final problem with the restricted-domain view is that it is unable to determine what kind of entity a REDO is.[37] The notion of a restricted domain of objects plays a crucial role in this view's attempt to explain what it is for an object X to exist. Consequently, this view should be able to explain what kind of entity a REDO is. For example, is a REDO just a collection of objects, or is it an entity which exists in its own right? Now, whatever this position suggests that a REDO is, a REDO is not such that a REDO's having the property of containing an object X and being [X]ish could explain what it is for X to exist. In other words, there are no entities called 'restricted domains of objects' which are such that their having the property of containing an object X and being [X]ish could explain what it is for an object X to exist.[38] No REDO is such that the existence of X can be identified with and reduced to the REDO's having the property of containing X and being [X]ish.

Chapter 5 concludes my systematic overview of the contemporary debate about the nature question. The upshot of this first part of my inquiry is that none of the discussed views can explain in a satisfactory manner what it is for an object that exists to exist. Their attempts to explain what it is for an object to exist are riddled with problems. This brings us to the second part of my book, which consists of two chapters.

In Chapter 6, I reconstruct three stages of Aquinas's answer to the nature question. Aquinas's first and preliminary answer to the nature question is that the existence of an object—a substance in his terminology—consists in that substance's having existence (*esse* in his terminology). But it is important to note that Aquinas does not claim that *esse* is a property or accidental form of a substance. The existence of a substance does not consist in the substance's having *esse* as a property or accidental form. Rather, in Aquinas's view, *esse* is primitive, i.e., it cannot be defined or analyzed any further. Nevertheless, this does not

---

37 My argumentative strategy in this section is inspired by Vallicella and follows him with regard to some distinctions but also significantly differs from his way of arguing, see Vallicella, *A Paradigm Theory of Existence*, 137–146.

38 For this formulation of the problem, I am indebted to ibid., 137.

mean that Aquinas regards a substance's *having esse* as primitive.[39] On the contrary, he entertains a sophisticated view about how this 'having' is supposed to work. Aquinas's first and preliminary explanation of how a substance's 'having' *esse* works is that a substance has *esse* by means of its essence or form (see 6.1).[40] Thus, what explains a substance's having *esse* is that a substance has an essence or form as a metaphysical part of it. A substance's essence or form plays a mediatory role: A substance has *esse* by means of an essence or form.

Now, this preliminary and provisional answer to the nature question needs a bit of unfolding. The first stage of Aquinas's explanation just points us in the right direction. It indicates only where we should start to find a satisfying answer to the nature question. What remains to be explained is how a substance has *esse* by means of an essence or form and how a substance is related to its *esse* if the substance's *esse* is not a property or accidental form of it. To answer these questions, Aquinas makes use of the idea of metaphysical participation. Accordingly, the second stage of Aquinas's explanation of what it is for a substance to exist is that a substance has *esse* by means of its essence or form in virtue of the participation of its essence or form in *esse* (see 6.2). Aquinas carefully distinguishes the way or mode *esse* is participated in by a substance from other ways or modes something is participated in by something else.

It is especially important to understand how he distinguishes the way or mode *esse* is participated in by the essence or form of a substance from the way an accidental form is participated in by a substance. What both ways or modes of participation have in common is that a substance participates in something else which is extrinsic to its essence or nature. But neither does *esse* determine the essence or form of the participating substance in any quidditative sense as some accidental forms of material substances do, nor does *esse* determine the substance in any accidental formal respect as every accident does. Rather, *esse* determines that which participates in it, namely the essence or form of this substance and consequently the substance as what it is essentially, with respect to its actuality.

By making use of the idea of participation, Aquinas's view helps us to see how a substance is related to its *esse*. Since a substance participates in *esse* as something which is not included in its essence or nature, the substance considered as what it is in virtue of its essence or form alone is a determinable with respect to its *esse*. Consequently, *esse* is a determinant of a substance which de-

---

39 Cf. Kerr, *Aquinas's Way to God*, 102.

40 I will explain below (see 6.1) why Aquinas distinguishes between the essence and the form of a material substance. For the moment, the distinction can be ignored. I am thankful to Jeremy Skrzypek for raising this worry.

termines the substance to be in actuality as what it is essentially. *Esse* is not an accidental form of a substance but rather the actuality of its essence or form, i.e., the actuality of a substance as what it is essentially.

But Aquinas is not done yet. What is still left unexplained is how metaphysical participation in general works, and particularly how the metaphysical participation works which is supposed to explain what it is for a substance to exist. To answer this question, Aquinas makes use of the idea of metaphysical composition. Whenever something is participated in, there must be something besides that which is participated in, namely, that which participates. In other words, metaphysical participation requires metaphysical composition, namely, a metaphysical composition of that which participates and that which is participated in, which are related as potentiality to actuality. In Aquinas's view, the participation of substance in *esse* is no exception to this general rule. Thus, the third and final stage of Aquinas's answer to the nature question is that the participation of a substance in *esse* results from its metaphysical composition of a determinable—its essence or form—and a determinant, namely, its *esse* (see 6.3).[41] In other words, the existence of a substance consists in its metaphysical composition of a determinable—the substance considered as what it is essentially in abstraction from its *esse* (*quod est*)—and a determinant, namely, its *esse* in virtue of which (*quo est*) the essence or form of this substance (and consequently the substance itself taken as what it is essentially) is actual.

In Chapter 7, I explain in detail why Aquinas's view can solve the problems which the views discussed in the first part cannot. The contemporary views discussed in the first part all agree that the existence of an object X consists in something *having a property F*. They disagree only about whether it is the object itself or something else which has F and about what F is. It is the acceptance of this premise which generates the identified problems.

I conclude that Aquinas's view makes three important contributions to the contemporary debate. First, Aquinas's view rejects a premise which so far has been taken for granted—that the existence of an object X consists in something having a property F—and it is this rejection which helps us to see why the contemporary debate about the nature question has reached an impasse.

If one accepts this premise, it is quite plausible to make use of one of the two standard metaphysical strategies to explain how something's having a property

---

41 I am thankful to Jonathan Nebel and Eleonore Stump who first suggested I think about Aquinas's view on the relation between a substance and its *esse* in terms of the relation between a determinable and a determinant.

F works.[42] According to a compositional explanatory strategy, what explains an object X's having a property or accidental form F is that F *inheres in* X as a metaphysical part of X. But the employment of this explanatory strategy generates the problems of the non-reductive and reductive-object view I have briefly sketched above. For example, the explanation of what it is for an object X to exist becomes viciously circular, because the inherence of a property F in X presupposes the existence of X in which it inheres.[43] Thus, these two views have the problems they have because they start with the premise that the existence of an object X consists in X's having a property F and employ a compositional strategy in order to explain how X's having F is supposed to work.

Now, if one rejects the idea that objects are metaphysically structured, i.e., that they have metaphysical parts, one cannot make use of a compositional strategy to explain an object X's having existence. What remains is the option to use a relational strategy to explain how something's having a property F works. According to this explanatory strategy, X does not have existence or any other property F in the sense that it is a metaphysical part of it. Rather, the explanation of X's having F is that X is related in a certain way to something else Y, which is extrinsic to X. [44]

So, F is not a monadic property of X but a dyadic or two-place property, i.e., a relation, which obtains between X and Y. For example, the explanation of the redness of a tomato is that the tomato as a concrete particular is related to the abstract object redness, which is external to it.[45] The redness of the tomato consists in a relation of instantiation, exemplification, or participation which obtains between the tomato as a concrete particular and redness as an abstract object which is external to the tomato.[46] Accordingly, the existence of an object X consists in a relation which obtains between X and something else Y which is extrinsic to X (e.g., a property or a domain of objects). For example, according to the property view, the existence of X consists in a relation of instantiation

42 For a brief but helpful sketch of these two explanatory strategies, see Kathrin Koslicki, *Form, Matter, Substance* (Oxford: Oxford University Press, 2018), 11–19.

43 It is precisely this problem which Barry Miller identifies as the central problem of a view which claims that the existence of an object X consists in X's having the property of existing and which he attempts to solve with his view, cf. Miller, *The Fullness of Being.*

44 Cf. Vallicella, *A Paradigm Theory of Existence*, 31.

45 Cf. Koslicki, *Form, Matter, Substance*, 15–16.

46 Cf. ibid., 16. For a more detailed exposition and defense of this explanatory strategy in terms of instantiation, see Peter van Inwagen, "Against Ontological Structure," in *The Problem of Universals in Contemporary Philosophy*, ed. Gabriele Galluzzo and Michael J. Loux (Cambridge: Cambridge University Press, 2015).

which obtains between a property Y which individuates X and X by which Y is instantiated.

The property view, the unrestricted-domain view, and the restricted-domain view have in common that they employ this kind of explanatory strategy. But it is not difficult to see why the employment of a relational explanatory strategy generates the problems which I have briefly sketched above. For example, if the existence of an object X consists in a relation which obtains between X and something else Y which is extrinsic to X, the problems of a vicious circularity and ontological inflationism arise because the obtainment of such a relation presupposes the existence of X and Y. Thus, these three views have the problems they have because they start with the premise that the existence of an object X consists in X's having a property F, and employ a relational strategy to explain how X's having F is supposed to work.

The second contribution of Aquinas's view to the contemporary debate is that it does not reformulate or develop an already existing and established position. Rather, his view introduces a new compositional explanatory strategy into the debate by disentangling the idea that the existence of an object X consists in X's having a property F from the idea that the existence of X consists in X's metaphysical composition. In this way, his view can integrate a valid insight of the compositional explanatory strategy without running into the problems which the non-reductive and reductive-object view are facing.

The third contribution of Aquinas's view is that it has the potential to integrate and synthesize the valid insights of the explanatory strategies it rejects. With a brief sketch of the core idea of Aquinas's answer to the strong ground question, I will show that Aquinas's position is also able to integrate a valid insight of the relational explanatory strategy. In Aquinas's view, a substance has the limited and caused *esse* it has but could also not have because it stands in a participatory relationship to unlimited and uncaused *esse*, i.e., subsistent *esse* itself, which is extrinsic to it.

Aquinas's combination of a compositional strategy to answer the nature question with a relational strategy to answer the strong ground question has a very interesting implication for his answer to the inventory question, i.e., his view on substances. On the one hand, a substance subsists. It is an *ens* which has *esse per se* and *in se* (see 6.1). On the other hand, a participatory relationship to subsistent *esse* itself, which is extrinsic to a substance, is constitutive of the metaphysical composition of this substance and consequently for its existence. So, in Aquinas's view, there is a constitutive relationality of that which subsists, and this relationality does not undermine the ontological status of substances as *entia* which subsist, i.e., which exist in their own right. Thus, according to the Thomistic synthesis, the structure of a substance as an *ens* by participation is indissolubly dyadic: It *subsists* as something which is constitutively *related* to something else, namely, to subsistent *esse* itself.[47]

Before I start with my inquiry, I need to address the question of how my book relates to the existing literature on this topic. To my knowledge, there is no monograph which brings into dialogue Aquinas's view on what it is to exist with the contemporary debate in the way I do. What makes my project unique is that it combines, first, a *systematic* reconstruction and criticism of the *contemporary* debate about the nature question with, second, a novel way to explain what the contribution of Aquinas's view to this debate is.

The *systematic* nature of my inquiry distinguishes it from books like John D. Caputo's *Heidegger and Aquinas*, Anthony Kenny's *Aquinas on Being*, G.E.M Anscombe's and Peter Geach's *Three Philosophers*, Norbert Bathen's *Thomistische Ontologie und Sprachanalyse*, or Giovanni Ventimiglia's *Aquinas after Frege*, which simply compare Aquinas's view with another view on what it is to exist.[48]

Furthermore, my methodological approach used to interpret and reconstruct what Aquinas's view is fundamentally distinguishes itself from the approach chosen by Ventimiglia, who follows the tradition initiated by Geach and continued by scholars such as Bathen and especially Kenny. These authors claim that we understand Aquinas's views better if we read them through the lens of modern authors like Gottlob Frege, i. e., by making use of the conceptual machinery provided by them. While this approach certainly has some merits, it has been widely criticized for being prone to distort Aquinas's views, especially when it comes to questions of existence. Instead, my method is guided by the idea that we understand Aquinas's views on a particular metaphysical issue better if we explain to the necessary extent the general metaphysical framework within which he works and the terminology that he uses to frame and to solve a particular metaphysical problem.

The focus on the *contemporary* debate distinguishes my book from works like Etienne Gilson's *Being and Some Philosophers*, Albert Keller's *Sein oder Existenz?*, or John X. Knasas's *Being and Some Twentieth-Century Thomists* and

---

47 Cf. William N. Clarke, *Explorations in Metaphysics: Being-God-Person* (Notre Dame, Indiana: University of Notre Dame Press, 1994), 108, 119–120; David L. Schindler, "Norris Clarke on Person, Being, and St. Thomas,"*Communio* 20, no. 3 (1993).

48 G.E.M. Anscombe and Peter T. Geach, *Three Philosophers: Aristotle, Aquinas, Frege* (Oxford: Basil Blackwell, 1961); Norbert Bathen, *Thomistische Ontologie und Sprachanalyse* (Freiburg im Breisgau; München: Verlag Karl Alber, 1988); John D. Caputo, *Heidegger and Aquinas: An Essay on Overcoming Metaphysics* (New York: Fordham University Press, 1982); Anthony Kenny, *Aquinas on Being* (Oxford; New York: Oxford University Press, 2002); Giovanni Ventimiglia, *Aquinas after Frege* (Cham, Switzerland: Palgrave Macmillan, 2020).

his recent *Thomistic Existentialism & Cosmological Reasoning*, which either compare Aquinas's view with older views on existence than I do (e.g., those of Descartes, Spinoza, Kant, and Hume), or are mostly concerned with debates between different Thomistic views on existence or *esse* which have been advanced in the 20th century.[49]

There are two categories of books which come very close to what I am doing in my work, because they address in a systematic way the contemporary debate about the nature question. The first category includes books like C.J.F. Williams's *What is Existence?*, Reinhardt Grossmann's *The Existence of the World*, Quentin Gibson's *The Existence Principle*, William F. Vallicella's *A Paradigm Theory of Existence*, Markus Gabriels' *Fields of Sense* and *Sinn und Existenz*, or Andreas Luckner's and Sebastian Ostritsch's *Existenz*.[50] What distinguishes these books from my own inquiry is that they lack an engagement with Aquinas's view on what it is to exist. Thus, they do not consider what Aquinas's view might contribute to the debate.

The second category includes books like the volume *Philosophie der Existenz* edited by Andreas Luckner and Sebastian Ostritsch, Barry Miller's *The Fullness of Being*, Gaven Kerr's *Aquinas's Way to God*, or *The Philosophy of Being in the Analytic, Continental, and Thomistic Traditions* by Joseph P. Li Vecchi, Frank Scalambrino, and David K. Kovacs.[51] These books offer a—more or less extensive—systematic treatment of the contemporary debate about what it is to exist and they try to show what Aquinas's view contributes to it.[52] My inquiry distinguishes itself from these books in two respects.

49 Étienne Gilson, *Being and Some Philosophers*, 2 ed. (Toronto: Pontifical Institute of Mediaeval Studies, 1952); Albert Keller, *Sein oder Existenz?* (Pullach bei München: Berchmanskolleg Verlag, 1968); John F. X. Knasas, *Being and Some Twentieth-Century Thomists* (New York: Fordham University Press, 2003); *Thomistic Existentialism & Cosmological Reasoning* (Washington, D.C.: The Catholic University of America Press, 2019).

50 Gabriel, *Fields of Sense*; *Sinn und Existenz*; Quentin Gibson, *The Existence Principle* (Dordrecht; Boston; London: Kluwer Academic Publishers, 1998); Grossmann, *The Existence of the World*; Luckner and Ostritsch, *Existenz*; Vallicella, *A Paradigm Theory of Existence*; C.J.F Williams, *What is Existence?* (Oxford: Clarendon Press, 1981).

51 Kerr, *Aquinas's Way to God*; Joseph P. Li Vecchi, Frank Scalambrino, and David K. Kovacs, *The Philosophy of Being in the Analytic, Continental, and Thomistic Traditions: Divergence and Dialogue* (London: Bloomsbury Academic, 2020); Andreas Luckner and Sebastian Ostritsch, eds., *Philosophie der Existenz: Aktuelle Beiträge von der Ontologie bis zur Ethik* (Berlin: J.B. Metzler, 2019); Miller, *The Fullness of Being*.

52 Miller does not explicitly claim that the view he presents is Aquinas's view. But it is pretty clear that he aims to defend a view he believes was Aquinas's view or at least something very close to it by making use of the conceptual machinery provided by contemporary analytic philosophy. So, perhaps more cautiously, I should say that he is defending an answer to the nature

First, I explain in a novel way what Aquinas has to say about the question of what it is for an object to exist. In particular, my book makes it clear that the idea of *something's having a property* plays no role in Aquinas's explanation of what it is for a substance to exist. The existence of a substance X does not consist in X's having existence or *esse* as a property *sui generis*. Rather, a substance is related to its *esse* as a determinable to a determinant. To my knowledge, no one has ever before tried to explain Aquinas's view in this way.

Second, due to this novel way of explaining what Aquinas's view actually is, I come to a completely different understanding of what the contribution of Aquinas's view to the contemporary debate about the nature question is. Contemporary interpreters of Aquinas such as Miller, Kerr, Turner C. Nevitt, or Andreas Schmidt argue that he entertains the view that *esse* is a property of an object that has it.[53] Consequently, they believe that Aquinas's view is a version of the view that the existence of an object X consists in X's having existence or *esse* as a property *sui generis*. So, according to them, Aquinas's view provides us at best with a way to solve the well-known problem that existence as a property cannot account for the existence of that of which it is a property, because having existence as a property presupposes the existence of a subject in which the property can inhere.[54]

---

question which he believes is Thomistic in spirit. But nothing important hinges on this, so I leave this subtlety aside.

53 Cf. Kerr, *Aquinas's Way to God*, 95 n.6, 100, 118, 143, 145–146, 192; Miller, *The Fullness of Being*, 162–169; Nevitt, "How to Be an Analytic Existential Thomist," 330, 342, 345; Andreas Schmidt, "Ipsum Esse subsistens. Thomas von Aquin und die existierende Existenz," in *Philosophie der Existenz: Aktuelle Beiträge von der Ontologie bis zur Ethik*, ed. Andreas Luckner and Sebastian Ostritsch (Berlin: J.B. Metzler, 2019), 1, 13. It should be noted that these authors are well aware that Aquinas does not claim that existence or *esse* is an accident or a property which inheres in a subject or is instantiated by an object that has it. So, they argue that they use the term 'property' in the broader sense of a characteristic or feature which a thing possesses to reconstruct Aquinas's view. The problem with this move is that it does not explain what is supposed to be explained, namely, how a substance is related to its *esse*. A notable exception is Miller who formulates a theory of the relation between a substance and the property of existence according to which existence is not a property which inheres in its subject but is only bound by it. I do not engage with Miller's rather complex and difficult theory because Miller still assumes that existence is a property of that which has it. I will show in this book, first, that this is not Aquinas's view and, second, that Aquinas has a better way to explain what the existence of an object consists in and how an object is related to its existence. Once the idea that existence is a property of that which has it is given up, there is no need to tell a complicated story about how having the property of existing can be responsible for the existence of its subject without presupposing an already existing subject.

54 Again, a large part of Miller's view is dedicated to solving this problem.

Instead, my interpretation and explanation of Aquinas's view show that his view is not a version of the non-reductive-object view. The contribution of Aquinas's view to the contemporary debate is more fundamental and far-reaching than that. It offers more than a new compositional strategy which solves the problem of explaining what it is for an object to exist. The consideration of Aquinas's view also shows why the whole contemporary debate has reached an impasse. Aquinas's view rejects what all contemporary views take for granted, namely, that the idea of something's having a property F must figure into an explanation of what it is for an object to exist.

Up to this point, I have explained what this book is about, I have given an overview of the whole argument, and I have clarified how my book relates to the existing literature on this topic. What remains to be done is justify the methodological approach of this inquiry by addressing some worries which might arise. A first worry might be concerned with the systematic nature of my reconstruction of the debate in the first part. Why have I chosen the five positions I deal with and not others?[55]

I have chosen these five views because they stand for the options which are available if one starts with the assumption that having a property F must play a role in the explanation of the existence of an object X. F can either be a property of X or a property of something else Y. If F is a property of X, it is either the property of existing or another property of X. So, in this sense, the first part exhausts the systematic options available. It presents in a systematic and exhaustive way what *types* of answers to the nature question, i.e., what types of theories of existence, can be formulated if one accepts the premise that the existence of X consists in something's having a property F.

The first part does not claim, of course, to present an exhaustive treatment of *tokens* of the different types of theories of existence. For example, instead of dealing with the view that the existence of an object X consists in X's having the property of being numerically identical with itself, I could have dealt with the view that the existence of X consists in X's having the property of being material, or the view that the existence of X consists in X's having the property of being living etc. And instead of dealing with the views that the existence of X consists in a UDO's having the property of containing X or a REDO's having the property of containing X, I could have dealt with views in which 'Y' stands not for a property or a domain of objects but something else.

However, such an exhaustive treatment of tokens of different types of answers to the nature question is neither desirable nor necessary, because many

---

55 I am thankful to Bruno Niederbacher and an anonymous referee for raising this worry.

of the problems I have identified concern the type and not the token, i.e., it does not matter which token is chosen to represent the type. For example, it does not matter what 'Y' stands for. Whatever Y is, it has to exist. So, any token of the type of answer to the nature question which argues that the existence of X consists in Y's having a property F generates the problems of an ontological inflationism and an infinite regress.

Nevertheless, my choice of tokens has not been entirely arbitrary. There are two criteria which guided my selection of the particular tokens of the types of views I deal with in the first part. The first criterion is strength. For example, as I will make clear in more detail below (see 2.1), if one wants to argue that the existence of an object X consists in X's having a property F and that F is not the property of existing, the most promising candidate of such a view is a version which maintains that the relevant property F is X's property of being numerically identical with itself.

The second criterion is newness. For example, I opted for Markus Gabriel's version of the view that the existence of an object X consists in a REDO's having the property of containing X because this theory of existence is a relatively new and innovative contribution to the ongoing debate which—at least in the analytically oriented philosophical community—has not received the attention it might deserve.

This brings me to a second methodological worry. In Chapters 2 and 5, I deal with particular authors who have defended the relevant view. Within these chapters, I do exegetical work and engage in depth with the writings of those authors. In contrast, Chapters 1, 3, and 4 are more schematic. Within these chapters, I only name authors who have defended the relevant views or content myself with stating that an author or several authors have been interpreted in the debate as embracing such a view. So, it could be objected that—since an explicit engagement with such authors is missing—it is difficult for the reader to verify whether the relevant author or authors have in fact entertained the view which is attributed to them.[56]

I would like to respond in two ways to this worry. First, it is true that the chapters of the first part are different in nature, but nothing problematic follows from it because I am cautious with the attribution of the views which I deal with in the relevant chapters. For example, in Chapters 3 and 4, I make it clear that I am not claiming that a person X has said p. Rather, I claim only that a person Y has said that X has said p, providing evidence for this more cautious claim in the footnotes. So, by consulting the footnotes the reader can verify whether what I

56 Thanks to Bruno Niederbacher and an anonymous referee for raising this worry.

claim that Y has said about what X has said about p is true. Thus, in these chapters, I take responsibility only for the fact that others have interpreted an author or authors in such a way that the view which I deal with in that chapter can be attributed to that author. I do not take responsibility for the accuracy of the interpretation.

Second, I do not engage in exegetical work in Chapters 1, 3, and 4 because this has already been done by other authors in other books. In short, there is simply no need to do so because in-depth treatments and interpretations of such authors with respect to what they have to say about the nature question are already available. For example, Chapters 4 and 5 of William Vallicella's *A Paradigm Theory of Existence* justify in great detail the attribution of the views I consider in Chapters 3 and 4 to authors like Frege, Quine, Russell, Frederic Sommers, and Nicolai Hartmann. Put positively, I do engage in exegetical work in Chapters 2 and 5 because this has not already been done and therefore contributes something new to the debate about the nature question.

Finally, there could be the methodological worry that it is somehow anachronistic to suppose that the views of a medieval author like Aquinas could contribute something to a debate in contemporary metaphysics.[57] As I explained above, I think it is anachronistic and methodologically problematic to suppose that, for example, the views of Frege and Aquinas can be brought into dialogue by somehow trying to 'translate' what Aquinas says about existence with the help of the Fregean conceptual machinery.

The method which I employ in this book is quite different. Methodologically, I assume that there are perennial philosophical questions, and that the nature question is one of them. As I explained right from the start, the question of what it is for an object to exist arises for any metaphysical theory which accepts that there are objects which exist. So, there is methodologically nothing problematic to bring into dialogue Aquinas's answer to the nature question with contemporary answers, if one takes into account that the relevant metaphysical frameworks within which the same perennial problem is framed and addressed are quite different.

This methodological caution also explains why I introduce and explain concepts such as 'form', 'essence', and 'participation' in Chapter 6. A reconstruction of Aquinas's answer to the question of what it is for an object to exist requires expounding, at least to a certain extent, the terminology and the general metaphysical framework within which he formulates his view.

---

57 I am thankful to an anonymous referee for making me aware of the need to address this issue.

As we will see below in detail (see 6.1), the nature question arises for Aquinas within an Aristotelian metaphysical framework in which forms and essences play an important role. The resources which this metaphysical framework provides—especially the Aristotelian idea of accounting for metaphysical composition in terms of actuality and potentiality—are used by Aquinas to answer the question of what it is for an object to exist. However, as I will explain in detail (see 6.2 and 6.3), Aquinas's view also makes use of the Platonic or (neo-)Platonic idea of participation to explain how a substance can have *esse* without it being the case that it is a property or accidental form of that substance. So, the metaphysical framework from within which he frames and solves the problem of explaining what it is for an object to exist is best understood as a synthesis of Aristotelian as well as Platonic or (neo-)Platonic elements.

By way of conclusion, it is important to note that I do not expect contemporary philosophers simply to accept Aquinas's metaphysical framework without further argument.[58] It is beyond the scope of this book to justify, for example, Aquinas's Aristotelian assumption that material objects have essences or forms. However, my book could be understood as an indirect argument in favor of a Thomistic metaphysics. The fact that the contemporary debate has reached an impasse shows that contemporary metaphysical frameworks, which are mostly reductive in nature and which try to stay clear from metaphysical ideas such as forms, essences, or participation, lack explanatory power when it comes to the fundamental metaphysical question of what it is for that which exists to exist.

Thus, there are two reasons which count in favor of the adoption of Aquinas's metaphysical framework. First, Aquinas's metaphysics provides the resources necessary to formulate a more satisfactory answer to the nature question. Second, the adoption of Aquinas's view on *esse* enables one to see and explain why the contemporary debate has reached an impasse and why the considered non-Thomistic views lack the resources to solve the problems they have in their attempts to answer the question of what it is for existents to exist.

---

58 Thanks to Godehard Brüntrup for making me aware of the need to address this worry.

Part I: The Contemporary Debate About the Nature Question

1 The Non-Reductive-Object View

1.1 The Core Idea of the View

Let us begin with the non-reductive-objective view, which gives the most straightforward and simple answer to the nature question: the existence of an object consists in that object's having the property of existing. This view has a long tradition within Western philosophy and can be traced back to Anselm of Canterbury's famous ontological proof for the existence of God. More recently, versions of this view have been defended by authors such as Francisco Berto, Colin McGinn, or Barry Miller.[1] What distinguishes this view from that of van Inwagen (dealt with in Chapter 2) is that it employs a non-reductive explanatory strategy. For an object to exist is to have the property *of existing*. Existence cannot be identified with and reduced to any other property an object has, e. g., the property of being identical with itself, as van Inwagen claims.[2]

This view involves three claims: First, existence is a property of an object that exists. Second, existence is a property *sui generis* of an object that exists, i. e., existence cannot be identified with and reduced to any other property of an object that exists. Finally, having the property of existence accounts for the existence of an object that exists. It is the conjunction of these three claims that I consider to be the core idea of the non-reductive-object approach.

Now, it is important to recall the limited scope of my book, the topic of which is the nature question, i. e., the question of how different views explain what it is for an entity that exists to exist. I am not asking about different responses to the question, 'What does the non-existence of a non-existing object consist in?'. Among those who maintain that the existence of an object X consists in X's having the property of existing, there are different answers to questions such as, 'Are there objects which do not exist?' or 'If there are objects that do not exist, how can one make sense of their non-existence?'. Nevertheless, whatever their various answers, they share the common idea that for an object to exist is to have a property, namely, the property of existing. And it is this common idea which I will criticize in the following.

---

1 Cf. Berto, *Existence as a Real Property*; McGinn, *Logical Properties*; Miller, *The Fullness of Being*. For other authors, see footnote 20 above.
2 Cf. van Inwagen, "McGinn on Existence," 55 – 56; *Metaphysics*, 300 – 301.

https://doi.org/10.1515/9783110979879-004

1.2 Three Problems

1.2.1 A Vicious Circularity

The first problem with the non-reductive-object view, i.e., the view that the existence of an object X consists in X's having the property of existing, is that its answer to the nature question is circular. An object can have or possess the property of existing only if it exists. The assertion that for an object to exist is to have the property of existing does not explain, but rather presupposes, the existence of that object.

What is supposed to be explained is what it is for an object X that exists to exist. According to the non-reductive-object view, the statement that X has the property of existing should do the explanatory work. Now, the problem is that this idea is dependent on the existence of X. X can have the property of existing only if X exists. Something that does not exist cannot *have* or *possess* the property of existing. Having the property of existing presupposes a pre-existing subject of that property.

The problem becomes clear if we ask in more general terms what it is for an object to have a property. Frequently, an object's having a property is expressed in terms of instantiation.[3] Properties are instantiated by objects while objects are never instantiated but only instantiate properties.[4] Thus, for an object to have a property is to instantiate that property. In other words, the explanation of an object's having a property is that the property is instantiated *by* the object. If this general account of property possession is applied to the case of an object's having or possessing the property of existing, it follows that an object has the property of existing in virtue of instantiating the property of existence. Thus, the explanation of an object X's having the property of existing is that existence is instantiated by X.

Now we see the circularity: Existence can be instantiated *by* X only if X exists. Consequently, X's having the property of existing cannot explain the existence of X because X's having the property of existing presupposes the existence of X by which the property of existence is instantiated.[5] It would be absurd to claim that existence can be instantiated by an object which does not have exis-

---

3 Cf. McGinn, *Logical Properties*, 15–16; Michael Nelson, "Existence," ed. Edward N. Zalta, Summer 2020 ed., *The Stanford Encyclopedia of Philosophy* (2020), https://plato.stanford.edu/archives/sum2020/entries/existence/.

4 Cf. "Existence."

5 Cf. Vallicella, Paradigm, 42–43.

tence.[6] But if the instantiation of the property of existence requires an object which has existence by which existence is instantiated, it follows that the view that the existence of an object X consists in X's having the property of existing is viciously circular. The view does not explain, but rather presupposes, what it is supposed to explain.

1.2.2 Existence Is Neither an Essential Nor an Accidental Property of an Object That Has It

The second problem with the view that the existence of an object X consists in X's having the property of existing concerns the nature of a property: If existence were a property of an object that exists, that existence would be either an essential or an accidental property of that object. But existence is neither an essential nor an accidental property of that object.

A more detailed formulation of the objection runs like this:[7]

(1) The existence of an object X consists in X's having the property of existing.
(2) Premise (1) implies that to exist is a property that determines its object.
(3) If a property determines its object, it is either an essential property or it is an accidental property of that object.
(4) Existence is not an accidental property of the object it determines.
(5) Existence is not an essential property of the object it determines.
Therefore, existence is not a property that determines its object.

Let us go through the different steps of the argument to evaluate its soundness. That (2) is an implication of (1) should be obvious. The whole point of the non-reductive-object view is that an object's having the property of existing makes a difference to X. Having the property of existing is supposed to explain in virtue of what an object X exists. But X's having the property of existing can explain the existence of X only if existence as a property determines the object that has that property—namely, with respect to the object's existence. In other words, the property of existing determines its object as an existing object. Consequently, (2) is an implication of (1).

---

6 Cf. Ibid., 45–48.

7 The general idea for this objection is inspired by Vallicella, *A Paradigm Theory of Existence*, 50. I am thankful to Eleonore Stump for helping me improve the formulation of this objection.

What about premise (3), which expresses a plausible view widely accepted by contemporary philosophers?[8] As such, a challenge to its truth is not likely to succeed. This leaves us with premises (4) and (5). While contemporary philosophers largely agree that the distinction between essential and accidental properties of an object is both exclusive and exhaustive (3), they dispute how best to characterize that distinction.

To evaluate the truth of premises (4) and (5), I do not need to take a stand on which way is the best way to characterize the distinction. Rather, I will take one standard way of characterizing the distinction and show that, based on this account, existence is neither an essential nor an accidental property of an object that exists. I will assume at the outset that using an alternate way of characterizing the distinction between essential and accidental properties will yield the same result.[9]

With the help of modal terms, contemporary philosophers have standardized ways of characterizing the distinction between essential and accidental properties, as illustrated by the following:[10]

Basic modal characterization (BMC):

> *P* is an essential property of an object *o* just in case it is necessary that *o* has *P*, whereas *P* is an accidental property of an object *o* just in case *o* has *P* but it is possible that *o* lacks *P*.

---

8 See, for example, Robertson Ishii, Teresa Atkins, and Philip Atkins, "Essential vs. Accidental Properties," ed. Edward N. Zalta, Winter 2020 ed., *The Stanford Encyclopedia of Philosophy* (2020), https://plato.stanford.edu/archives/win2020/entries/essential-accidental/; Vallicella, *A Paradigm Theory of Existence*, 49–50.

9 In Chapter 6, I will support this claim by showing that with Aquinas's way of characterizing the distinction between essential and accidental properties (which is significantly different from the account I will use in this section) we get the same result—namely, that existence is neither an essential nor an accidental property of an object that exists. Actually, the way Aquinas characterizes the distinction between essential and accidental properties seems to me quite close to an important contemporary alternate way of characterizing the distinction which has been proposed, for example, by Kit Fine ("Essence and Modality: The Second *Philosophical Perspectives* Lecture," *Philosophical Perspectives* 8, Logic and Language, 1994). So, in the course of this inquiry, I will show that based on two rather prominent ways of characterizing the distinction between essential and accidental properties, existence is neither an essential nor an accidental property of an object that has it.

10 The formulations are taken over from Ishii, Atkins, and Atkins, "Essential vs. Accidental Properties."

Possible world characterization (PWC):

> *P* is an essential property of an object *o* just in case *o* has *P* in all possible worlds, whereas *P* is an accidental property of an object *o* just in case *o* has *P* but there is a possible world in which *o* lacks *P*.

Existence-conditioned modal characterization (ECMC):

> *P* is an essential property of an object *o* just in case it is necessary that *o* has *P* if *o* exists, whereas *P* is an accidental property of an object *o* just in case *o* has *P* but it is possible that *o* lacks *P* and yet exists.

Based on these modal characterizations of the distinction between essential and accidental properties, we can see why existence is neither an essential nor an accidental property of an object that exists. Let us begin with premise (4) which states that existence is not an accidental property of the object it determines. According to the BMC, a property is an accidental property of an object if an object has the property, but it is possible that it could lack this property. For example, Angela Merkel happens to have the property of being the *Bundeskanzlerin* of Germany in spring 2021, but it is possible that she could lack this property. In the year 2022, she indeed lacks this property but she still exists. By contrast, being a human being is an essential property of Angela Merkel because she necessarily has this property. It is not possible for her to lack this property at any point in time.

Now, according to the BMC, existence is not an accidental property of an object because it is not possible for an object which has the property of existing to lack the property of existence. If an object which has existence ceases to have existence, it is no more. In other words, the loss of any accidental property, according to the BMC, does not terminate the very existence of the object which has it. But the loss of the property of existence would. Consequently, existence is not an accidental property according to the BMC.

This implicit presupposition of existence in the characterization of the distinction is made explicit by the existence-conditioned modal characterization of the distinction between essential and accidental properties. According to the ECMC, a property is an accidental property of an object just in case this object has that property, but it is possible that the object lacks the property and yet exists.[11] I think this is a good characterization for what an accidental property of

11 Such an existence-conditioned modal characterization of the difference between essential and accidental properties is widely in use. See, for example, Kris McDaniel, *This Is Metaphysics:*

an object is. But it is obvious that according to this characterization existence is not an accidental property of an object. It is simply not possible that an object lacks existence and yet exists.

This leaves us with the PWC, according to which a property is an accidental property of an object just in case the object happens to have the property but there is a possible world in which the object lacks the property. In contrast to the BMC, it seems that the PWC can make sense of the claim that existence is an accidental property of an object. For example, existence is an accidental property of Angela Merkel because she has existence in this actual world, but there is a possible world in which she lacks existence and does not exist.[12] This possible world is simply a world in which Angela Merkel does not exist. So, since she does not exist in all possible worlds, existence is not an essential, but just an accidental, property of hers.

But this way of characterizing existence as an accidental property of an object generates a serious problem for the characterization of the essential properties of an object. The PWC is forced to accept that properties which are most likely essential properties of an object are accidental properties of that object.[13] In other words, the PWC of existence as an accidental property of an object produces erroneous results when it comes to the characterization of the essential properties of an object.

To see why, let us consider the following example.[14] It should be rather uncontroversial that however we want to characterize the distinction between essential and accidental properties, we do not want to come up with the result that being a human being is an accidental property of Angela Merkel or of

---

An Introduction (Hoboken, NJ: Wiley-Blackwell, 2020), 256; Vallicella, *A Paradigm Theory of Existence*, 49.

12 It is important to note that the PWC is only able to make sense of the claim that existence is an accidental property of an object that has it if possible worlds are not themselves actual worlds, i.e., maximal mereological sums of *concreta*, cf. "Existence: Two Dogmas of Analysis," 66. Otherwise, existence is not an accidental property of an object because there is no possible world in which it is possible for an object to exist and to lack existence. Thus, for the sake of argument, I assume that possible worlds are not themselves actual worlds because otherwise the argument that existence is an accidental property would be a non-starter right from the beginning in the PWC as well. I will leave it open what exactly the ontological status of possible worlds is if they are not actual worlds. Rather, I treat talk about possible worlds in this context simply as a useful heuristic tool to characterize the difference between essential and accidental properties. For a similar use of the talk about possible worlds, see Eleonore Stump, *Aquinas* (London: Routledge, 2003), 110–115; Vallicella, "Existence: Two Dogmas of Analysis," 66.

13 Here I follow Ishii, Atkins, and Atkins, "Essential vs. Accidental Properties."

14 My formulation of this example is inspired by ibid.

any other human being. But this is precisely what follows from the PWC of existence as an accidental property of an object that exists. If existence is an accidental property of Angela Merkel, she has existence in this actual world but there is at least one possible world in which she lacks existence and does not exist. This possible world is simply a world in which Angela Merkel does not exist. But if she does not exist in that possible world, she does not have the property of being a human being in that possible world. She is not a human being in that possible world for the simple reason that she does not exist in that world. And now we see the problem. The PWC is forced to state that being a human being is not an essential, but an accidental, property of Angela Merkel because she does not have this property in all possible worlds. She does not have it in all possible worlds because existence is an accidental property of Angela Merkel, which means that there is at least one possible world in which she does not exist and consequently does not have the property of being a human being in that world.

To avoid this absurd result, the PWC of essential properties needs to be improved. An effective and elegant way to do so is to add an existence condition:[15]

Modified possible world characterization (MPWC):

> *P* is an essential property of an object *o* just in case *o* has *P* in all possible worlds in which *o* exists, whereas *P* is an accidental property of an object *o* just in case *o* has *P* but there is a possible world in which *o* lacks *P* and yet exists.

With the addition of the existence condition, the MPWC can describe essential properties in such a way that we get the correct results. For example, being a human being is an essential property of Angela Merkel because it is a property which Angela Merkel has in all the possible worlds in which she exists. There is no possible world in which Angela Merkel exists and lacks the property of being a human being. In a similar way, the MPWC yields sound results when it comes to the characterization of accidental properties. For example, being *Bundeskanzlerin* is an accidental property of Merkel because she happens to have this property in this actual world in spring 2021, but there is a possible world in which she lacks this property and yet exists, e.g., a possible world in which she exists but is not *Bundeskanzlerin* because the fall of the Berlin Wall never took place in that possible world.

The only drawback with the MPWC is that it does not categorize existence as an accidental property of an object. In contrast to the PWC, it is therefore unable to make sense of the claim that existence is an accidental property of an object

15 The formulation is taken over from ibid.

that exists. The introduction of the existence condition forces the MPWC to give up the claim that existence is an accidental property because there is no possible world in which an object lacks existence and yet exists.

To sum up: The PWC of the distinction between essential and accidental properties can make sense of the claim that existence is an accidental property of objects. But that way of characterizing existence as an accidental property leads to a serious problem in characterizing the essential properties of an object. This problem can be remedied by adding an existence condition, which provides a good characterization of essential and accidental properties. But in the MPWC's more satisfactory description of the distinction between essential and accidental properties, existence is no longer an accidental property of an object that exists.

We have yet to deal with premise (5), which states that existence is not an essential property of the object it determines. Let us go through the different modal characterizations of an essential property to see why this premise is true. According to the BMC of essential and accidental properties, existence is an essential property of an object that exists just in case it is necessary that the object has existence.

This claim is obviously false. To say that it is necessary that an object has existence is equivalent to saying that the object must have existence, i. e., that it could not lack existence.[16] Thus, the claim that existence is an essential property of an object that has existence amounts to the claim that the objects which have existence could not lack existence. In other words, it is not possible, for objects which have existence and exist, not to have existence and not to exist. Now, this claim is belied by our experience. For example, the apple I ate for lunch had existence before my lunch but not afterwards. It no longer exists after my lunch. It follows that it was possible for the apple which had existence not to have existence. This simple experience about the contingency of the apple can be generalized: in our experience, any existing object with which we are familiar could possibly not exist. It follows that existence is not an essential property according to the BMC of essential properties.

The same point can made from a slightly different angle.[17] If existence is an essential property of any object which exists, we cannot make sense of our overwhelming experience that objects which exist at one point of time t cease to exist at a later point of time t+1. Assuming that existence is a property of an object that exists, and if we account for the fact that an object can cease to exist, existence

16 I rely here on the characterization of 'necessity' and 'possibility' given by ibid.

17 The following argument is inspired by Vallicella, *A Paradigm Theory of Existence*, 50.

must be a property which can be lost. But if existence is an essential property of that which exists, existence cannot be lost because it is not possible for an existing object to lose or to fail to have a property which is essential to it. Therefore, existence is not an essential property of that which exists.

What about the PWC of essential and accidental properties? According to the PWC, existence is not an essential property of objects that exist because it is not the case that objects which have existence (in our actual world), have existence in all possible worlds. Objects which have existence and exist are not necessary beings which exist in all possible worlds.

This leaves us with the ECMC and the MPWC of the distinction between essential and accidental properties. What is common to both characterizations is that they introduce an existence condition to explain what an essential property of an object is. Like in the case of accidental properties, I think that the ECMC is a good characterization of what an essential property of an object is. Or, to be more precise, I think it successfully formulates at least a necessary condition that a property must satisfy to be an essential property of an object. Essential properties of an objects are those properties of an object which that object has necessarily if it exists.

Now, two problems arise if a proponent of the non-reductive-object approach uses the ECMC or the MPWC of essential properties to defend the claim that existence is an essential property of an object it determines. First, if existence is an essential property of an object, then an object necessarily has this property if it exists. However, this is wrong for the reasons we have seen above. The objects which we are familiar with in experience fulfill the existence condition, but this does not imply that they exist necessarily.

It is important not to get confused here with *de dicto* and *de re* ascriptions of necessity. A proponent of the non-reductive-object view is forced to conclude that objects have existence necessarily if existence is an essential property of objects that exist because they are talking about modal properties *of objects* and not of sentences. Thus, they cannot argue that the claim that an object has existence essentially (when it exists) amounts to the trivial claims '*Necessarily*, an object has existence, if it exists' or '*Necessarily*, an object has existence in all possible worlds in which it exists'. Rather, since they are talking about modal properties *of objects*, the claim that an object has existence essentially, if it exists, amounts to the strong metaphysical claims 'An object that has existence, if it exists, has existence *necessarily*' or 'An object that has existence in all possible worlds in which it exists has existence *necessarily*'. Thus, if a proponent of the non-reductive-object approach uses the ECMC or the MPWC of essential properties to defend the claim that existence is an essential property of an object it determines they are forced to draw the wrong conclusion that everything that exists, if it exists, exists necessarily.[18]

Second, a proponent of the non-reductive-object approach cannot use the ECMC or the MPWC of essential properties to defend the claim that existence is an essential property of an object it determines, because this would make their answer to the nature question inexorably circular: that which is supposed to explain the existence of an object—an object's having the property of existing as an essential property—appeals to the existence of an object to explain what an essential property of that object is.[19] In other words, since a reference to the existence of an object is used to explain what an essential property of an object is, an explanation which states that an object has existence because it is an essential property of it is viciously circular. Thus, the ECMC and the MPWC of essential properties are of no use in formulating an argument in favor of the claim that existence is an essential property of an object it determines.

Now, if a property which determines its object is either an essential or an accidental property of that object (3), but existence is neither an accidental (4) nor an essential property of the object it determines (5), it follows that premise (2) is false. Existence is not a property which determines its object. However, since the truth of (2) is an implication of the view that the existence of an object X consists in X's having the property of existing, it follows that the existence of X does not consist in X's having the property of existing. The existence of X does not consist in X's having the property of existing because existence is not a property which determines its object. If existence is not a property which determines its object in any way, having this property cannot account for the existence of that object. In other words, if an object X's having the property of existing is supposed to explain the existence of X, existence must be a property which determines X in some way.

1.2.3 This View Does Not Explain What It Is Supposed to Explain

A third problem of the non-reductive-object view comes to light if we consider what options there are for the view to rebut the objection I have reconstructed in the previous section. I have argued in detail that it looks unpromising to chal-

18 I am thankful to Timothy Pawl for making me aware of the need to clarify this point.

19 Like in the case of accidental properties, contemporary philosophers often explain what an essential property of an object is by introducing an existence condition. For example, according to McDaniel, a property is an essential property of that object, iff it is not possible for that object to exist without having that property, McDaniel, *This Is Metaphysics*, 258.

lenge either the truth of the premise that existence is not an accidental property of the object it determines (4), or the truth of the premise that existence is not an essential property of the object it determines (5). This leaves this view with two options. Either it attacks the premise that the claim that existence is a property which determines its object is an implication of the non-reductive-object view (2), or it attacks the premise which states that if a property determines its object, it is either an essential or an accidental property of that object (3).

Giving up (2) doesn't look very promising. If existence is a property which does not determine its object in any way, it follows that having the property of existing does not determine its object with respect to its existence. But if having existence as a property does not determine its object with respect to its existence, the non-reductive-object view fails to explain what it is for an object that exists to exist. Thus, giving up (2) is no option for this view because it undermines its explanatory power. By claiming that the existence of an object X consists in X's having existence as a property which does not determine X in any way—not even with respect to its existence—, the non-reductive-object view no longer provides an answer to the nature question.

The only remaining option for the non-reductive-object view is to attack premise (3). The attack runs as follows: Premise (2) is accepted. Existence is a property which determines its object, namely, it determines its object with respect to its existence. Consequently, the non-reductive-object view has explanatory power. An object X's having the property of existing explains the existence of X because existence is a property which determines the object which has it to be an existing object. Now, to solve the problem that existence is neither an essential nor accidental property of the object it determines (3), the non-reductive-object view rejects the idea that this distinction is both exclusive and exhaustive.

Rather, existence is a property *sui generis.*[20] Existence is a property *sui generis* because it determines its object, but it does not determine its object as an essential or accidental property determines its object. By having existence as a property *sui generis*, an object is neither determined in an essential nor in an accidental way but only with respect to its existence. It is this characteristic which makes the property of existence unique as a property which determines its object. To sum up: To rebut the objection that existence as a property which determines its object is neither an essential nor an accidental property of it, the non-

20 The strategy to declare that existence is a property *sui generis* to solve certain problems is not only used by views which argue that the existence of an object X exists in X's having the property of existing, but also by views which argue that the existence of an object X consists in something else Y's having a property. For example, see Gabriel, *Sinn und Existenz*, 78, 95–96; McGinn, *Logical Properties*, 44, 50; Miller, *The Fullness of Being*, 84–111.

reductive-object view must be amended to claim that the existence of an object X consists in X's having a property *sui generis*, namely, the property of existing.

But this amended view is confronted with a third problem, namely, that it does not explain what it is supposed to explain. What it is supposed to explain is what is for an object that exists to exist, i.e., what the existence of an object consists in. The answer of the original view was that for an object to exist is to have the property of existing. Existence is a property which determines its object with respect to its existence. So, the idea that an object is related to its existence in the way an object is related to a property which determines it is supposed to do the explanatory work.

Now, pressed by the objection that this explanation does not work because existence does not determine its object like an accidental or essential property, the non-reductive-object view is amended to claim that existence is a property *sui generis* of an object which determines its object with respect to its existence. But this move is unsatisfying because by qualifying existence as a property *sui generis* the non-reductive-object view leaves us where we started. The explanation that an object exists in virtue of having a property *sui generis* which determines it to be an existing object does not explain anything. It does not explain anything because it is left unexplained how an object is related to existence as a property *sui generis* of it and how existence as a property *sui generis* determines its object with respect to its existence or actuality.

To claim that existence as a property *sui generis* determines its object with respect to its existence explains nothing, because it remains unexplained how a property *sui generis* determines its object. Something which needs to be explained, namely, how an object is related to its existence and how existence as a property determines its object, is explained with something which itself needs to be explained, namely, how an object is related to existence as a property *sui generis* and how existence as a property *sui generis* determines its object. Thus, unless it is explained how a property *sui generis* determines its object and how such a property *sui generis* relates to an object which has it, nothing is explained by the view that for an object to exist is to have existence as a property *sui generis*.[21]

---

21 In addition to this worry, the non-reductive-object view is vulnerable to the following objection: According to this view, it seems that existence is a *trivial* property, i.e., that it is explanatorily superfluous. But it seems that questions about the existence of objects that exist are not trivial. For example, it seems to be a meaningful and substantive question to ask why an object which could lack existence has existence. Now, if the answer to this question is that an object that exists exists in virtue of having the property of existing, it is a meaningful and substantive question to ask how this 'having' is supposed to work. But if those questions are meaningful and

To sum up: The view that the existence of an object consists in that object's having existence as a property *sui generis* does not explain what it is supposed to explain. It is intelligible what it is for an object to have an accidental or an essential property. However, it is unintelligible what it is for an object to have a property *sui generis* which is neither essentially nor accidentally related to an object which has it. Thus, the attempt to explain the existence of an object by making use of the idea that existence is a property *sui generis* of an object that exists is unsatisfactory, because it does not explain how that which exists is related to existence as its property. In other words, it is not explained how existence as a property *sui generis* determines the object which has it with respect to its existence, if existence as a property *sui generis* determines its object neither essentially nor accidentally as any other property which determines an object that has it.

substantive, we need a non-trivial answer to the nature question. I am thankful to James Dominic Rooney for this insight.

2 The Reductive-Object View

The view which I have criticized in the previous chapter employed a *non-reductive* strategy to explain what it is for an object to exist. The existence of an object consists in that object's having the property of existing. Existence is a property *sui generis* of an object that exists which cannot be identified with and reduced to any other property of that object. In the remaining four chapters of this first part of my book, I reconstruct and criticize four views which employ a *reductive* explanatory strategy.

The employed explanatory strategy is reductive in nature because these views try to identify and reduce the existence of an object with and to something's having a property other than existence. But what is that something and what is the property of that something to which the existence of an object is reduced to? This is where the four views part company. The view which I consider in this chapter argues that the existence of an object X consists in that *object X's* having a property, namely, the property of being numerically identical with itself.[1] In contrast, the other three views maintain that the existence of an object X consists in *something else Y's* having a property.

2.1 The Core Idea of Peter van Inwagen's Version of This View

2.1.1 A General Problem with This View

Two claims are constitutive of the reductive-object view I am dealing with in this chapter. First, this view maintains that the existence of an object X consists in that *object X's* having a property. In this respect, the reductive-object view agrees with the non-reductive-object view considered in the previous chapter and disagrees with the other three views I will engage with in the three remaining chapters of this first part. According to the latter three views, the existence of an object X does not consist in X's having a property but in *something else Y's* having a property.

Second, the reductive-object view holds that although the existence of an object X consists in X's having a property, the relevant property is not existence.

1 In what follows, I leave out 'numerically' because I am not dealing with other conceptions of identity. Thus, 'being identical with itself' always means 'being numerically identical with itself' in this chapter.

https://doi.org/10.1515/9783110979879-005

There is no such thing as the existing objects' property *sui generis* of existing. Rather, the existence of an object X consists in X's having a property other than existence. The existence of an object X can be identified with and reduced to X's having a property other than existence. It is this reductionist claim which distinguishes the reductive-object view from the non-reductive-object view I presented in the previous chapter.

So far, I have characterized the reductive-object view in a general and neutral way because I have left it open what the property is which is supposed to explain the existence of an object that exists. For example, the view that the existence of an object X consists in X's having the property of being material and the view that the existence of an object X consists in X's having the property of being perceived are both variants of the general view that the existence of an object X consists in X's having a property (other than existence).[2]

Now, an important problem with all variants of the general view is that their reductive explanation of what it is for an object to exist does not cover all cases of objects that exist. In other words, as soon as any variant of the reductive-object view tries to specify what the property of an object is which is supposed to account for the existence of an object, it becomes evident that the view does not work for all cases of existing objects.[3] Consider, for example, the view that the existence of an object X consists in X's having the property of being material. At best, this view can explain the existence of material objects. Material objects exists in virtue of having the property of being material.

But are all existing objects material objects? Well, over the centuries many philosophers have advanced good reasons for believing that immaterial objects like God, separate substances as angels or disembodied souls, abstract universals, or minds exist. Now, if it is true that some of the objects which exist are immaterial objects, the view that the existence of an object X consists in X's having the property of being material is false because immaterial objects do not have the property of being material but do exist. The existence of any immaterial object shows that the reductive attempt to identify and reduce existence with and to an object's property of being material does not work.

Now, a defender of the view that the existence of an object consists in that object's having the property of being material can of course try to rebut this objection by denying that any immaterial object exists. The objection is rebutted if all existing objects are material objects. If all existing objects are material objects, there are no objects whose existence cannot be explained by this view.

---

2 Cf. van Inwagen, *Metaphysics*, 293.

3 In what follows, I am indebted to ibid., 292–296.

Thus, based on the premise that such a materialist worldview is true, the materialist explanation which is offered works, of course, for all cases of existing objects. In other words, if the materialist answer to the inventory question, i.e., the question 'What kinds or types of objects exist?' is the correct one, the materialist answer to the nature question is the correct one.

However, why should one accept a materialist worldview? The response cannot be that a materialist worldview is true because only material objects exist. Such a response is viciously circular. Consequently, independent reasons must be given for the truth of a materialist worldview and the claim that only material objects exist. In other words, the rebuttal of this objection (that the view that the existence of an object X consists in X's having the property of being material does not work because it cannot explain what it is for immaterial objects to exist) requires defending a materialist worldview (according to which all existing objects are material objects) on independent grounds. But this requirement makes the view extremely vulnerable. One sound argument for the existence of an immaterial object suffices to demonstrate that the existence of an object does not consist in an object's having the property of being material.

What I have said so far seems to apply to every other variant of the general view that the existence of an object X consists in X's having a property (other than existence). Whatever property existence is identified with and reduced to, the respective view is vulnerable to counterexamples of objects which lack the relevant property but do exist.

2.1.2 Van Inwagen's Attempt to Solve This Problem

There are two ways to solve this difficulty faced by the reductive-object view. The first strategy is to give up the claim that the existence of an object consists in that object's having a property. According to this strategy, existence is not only not a property *sui generis* of objects which exist, but not a property of existing objects at all.[4] This might explain why every attempt to identify and reduce the existence of an object with and to a property (other than existence) of that object must fail. Such attempts must fail for the simple reason that the existence of an object does not consist in that object's having a property. Existence is simply not a property which an object that exists has. The rejection of the claim that the existence of an object consists in that object's having a property is constitutive of the three views I deal with in the next three chapters.

4 Cf. ibid., 296–298.

But before I get to these views, I would like to consider a second strategy to solve the described difficulty of the reductive-object view. According to this strategy, the reductive-object view can be rescued through the identification and reduction of an object's existence with and to a property of an object which no existing object lacks. In other words, a property must be identified which allows no counterexamples of objects that exist but lack the relevant property.

Peter van Inwagen has recently suggested that a property which meets this criterion is the property of being numerically identical with oneself. For an object to exist is to be identical with itself.[5] Thus, van Inwagen defends a version of the general view that the existence of an object consists in that object's having a property according to which the existence of an object consists in that object's having the property of being identical with itself. Existence can be identified with and reduced to an object's property of being identical with itself.[6]

At first sight, the identification and reduction of an object's existence with and to that object's property of being identical with itself looks like a clever and promising move for the following reason: In the previous section, we have seen that a decisive objection against any variant of the general view that the existence of an object consist in that object's having a property is that it is vulnerable to counterexamples. Every variant of the general view is confronted with the problem that it seems that there are some objects which exist and yet lack the property with which the existence of an object is identified with and reduced to.

Now, van Inwagen's version of the general view can solve this problem. His view is not vulnerable to counterexamples because every existing object has the property of being numerically identical with itself. It would be absurd to claim that an object lacks the property of being identical with itself and yet exists. *What* object or kind of object would that be? So, it seems that an object's property of being identical with itself is the ideal candidate for a property with which the existence of an object can be identified with and reduced to: On the one hand, self-identity is a property which is general enough. Every existing object has the property to be identical with itself. On the other hand, this property is not a property *sui generis* of an object.

5 Cf. van Inwagen, "McGinn on Existence," 55–56; *Metaphysics*, 300–301.
6 Cf. *Metaphysics*, 301.

2.1.3 Van Inwagen's Motivation

Before I expound three serious problems with van Inwagen's version of the general view, a few more things must be said about his motivation to adopt it. Van Inwagen adopts his version of the reductive-object view, i.e., the view that the existence of an object consists in that object's having the property of being identical with itself, to rebut an objection against a semantic claim he is committed to. Van Inwagen is committed to the semantic claim that the term 'exists' is univocal, i.e., that it has the same meaning when it is applied to objects in different logical or ontological or metaphysical categories.[7] According to this semantic claim, 'exists' has the same meaning, i.e., it is used in one and the same sense, when it is applied to, for example, 'tangible object', 'mental object', or 'abstract object'.[8]

Now, the question is of course why he wants to defend this semantic claim. Van Inwagen appears to be convinced that he must defend this semantic thesis to undermine intuitions deriving from the way we use 'exists' in our ordinary ways of talking which in turn motivate us to adopt the ontological or metaphysical view that objects exist in different ways. In other words, van Inwagen believes that he must defend the univocity of 'exists' to undermine the idea that 'exists' has different meanings when it is applied to objects in different logical or metaphysical categories, because the latter idea motivates a metaphysical view which is called 'ontological pluralism' in contemporary analytic metaphysics.

According to ontological pluralism, there are ways or modes of existing.[9] For example, according to one version of ontological pluralism, substances and accidents exist but they do not exist in the same way. Substances *subsist* and accidents *inhere in* substances. So, there are at least two modes or ways of existing: Subsistence and inherence. Thus, an important implication of any version of ontological pluralism is that there is more to be said about the existence of an object than the mere fact *that* a thing exists. Objects that exist do not simply exist but they exist in different ways, i.e., they have different modes or ways of existing. Consequently, the opposite view to ontological pluralism is called 'ontolog-

---

7 Cf. "Being, Existence, and Ontological Commitment," 482. Van Inwagen states in footnote 19 of that same page that the meaning of the expression 'logical or metaphysical category' is far from clear but that its precise meaning is not relevant to the question whether objects in different 'categories' exist in different senses of 'exist'.

8 Cf. ibid.

9 Cf. McDaniel, *The Fragmentation of Being*, 4; Turner, "Ontological Pluralism," 5.

ical monism' because according to this view there is nothing more to be said about the existence of an object other than that it exists.[10]

Now, any version of ontological pluralism challenges van Inwagen's attempt to identify and reduce the existence of an object X with and to X's having the property of being identical with itself, because the idea that there are modes of existing, i.e., the idea that objects have existence and exist in different ways, cannot be adequately captured by the existential quantifier of formal logic.[11] In other words, the view that existing objects have existence and exist in different ways cannot be expressed with a standard first-order predicate logic which operates with a single existential quantifier.

Look at it this way. Van Inwagen considers his view to be a version of a view which has been entertained by thinkers like Gottlob Frege, Bertrand Russell, and W.V.O. Quine.[12] A central characteristic of the latter view—I will present and criticize it in more detail in the next chapter—is that the existence of an object is identified with and reduced to a property's property of being instantiated. Consequently, this view is challenged by any version of ontological pluralism because it cannot accommodate the idea that there are ways or modes of *existing* for the simple reason that there are no ways or modes of *instantiation*. Either a property is instantiated, or it is not. It does not make sense to ask 'Ok, a property is instantiated. But in which way?'.[13]

In an analogous way, van Inwagen's view that the existence of an object consists in that object's having the property of being identical with itself is challenged by any version of ontological pluralism because it cannot accommodate the idea that there are ways or modes of existing, i.e., the idea that objects which exist have existence in different ways. Either an object has the property of being identical with itself or it doesn't. Nothing more can be said. It makes no sense to say that objects which are identical with themselves have the property of being identical with themselves in different ways. To sum up: The view that there are modes or ways of existing undermines a claim to which van Inwagen is committed, namely, that all there is to say about the existence of an object can be adequately captured and expressed by the single existential quantifier of formal logic.

---

10 Cf. McDaniel, *The Fragmentation of Being*, 140–141.

11 Cf. van Inwagen, "Being, Existence, and Ontological Commitment," 492–499; Ross P. Cameron, "Critical Study of Kris McDaniel's *The Fragmentation of Being*," *Res Philosophica* 95, no. 4 (2018): 785–787.

12 Cf. van Inwagen, "Being, Existence, and Ontological Commitment."; "McGinn on Existence," 51, 55–56; *Metaphysics*, 296–301.

13 Cf. Vallicella, "Existence: Two Dogmas of Analysis," 59.

Now, to block the path to any version of ontological pluralism, van Inwagen tries to defend the semantic claim that 'exists' is univocal. In this way, he hopes to undermine the semantic view which is an important motivation for the adoption of a corresponding ontological pluralism, namely, the view that 'exists' has different meanings when it is applied to objects in different categories.[14] Van Inwagen's main argument for the univocity of 'exists', and against the intuition that the term 'exists' has different meanings when it is applied to objects in different categories, can be reconstructed in the following way:[15]

(1) Number-words such as 'six' or 'forty-three' do not mean different things when they are used to count objects of different sorts, i.e., objects in different logical or ontological or metaphysical categories. For example, if the number word 'thirteen' is used for counting the epics written by a person A and is used for counting the cats a person B possesses, 'thirteen' is used in the same sense in both cases.
(2) Existence is closely allied to number, i.e., existential statements are logically equivalent to statements of number. For example, the following statements are logically equivalent:
    a) 'Unicorns do not exist' is logically equivalent to 'The number of unicorns is 0'
    b) 'Horses exist' is logically equivalent to 'The number of horses is 1 or more'
    c) 'Angels, or ideas, or prime numbers exist' is logically equivalent to 'The number of angels, or of ideas, or of prime numbers is greater than 0'

Therefore, the term 'exists' does not have different meanings when it is applied to objects in different logical or ontological or metaphysical categories.

The structure of this 'univocity of number' argument is rather simple:[16] Number-words such as 'one' or 'two' etc. are univocal (1). There is an intimate connection between existence and number, such that existential statements are logically equivalent to statements of number (2). Therefore, the term 'exists' is univocal when it is applied to objects in different logical or ontological or metaphysical categories.

I think premise (1) of this argument for the univocity of 'exists' is rather unproblematic and uncontroversial. The problem is premise (2), i.e., the Fregean doctrine of the relation between number and existence, which is usually sum-

14 Cf. van Inwagen, "Being, Existence, and Ontological Commitment," 482–492.
15 Cf. ibid., 482.
16 Cf. ibid., 492 n. 33.

marized in the claim that the affirmation of existence is a denial of the number 0.[17] To be more precise, the problem is not the Fregean doctrine as such but van Inwagen's interpretation of it, which deviates from the traditional interpretation of this doctrine. To see what the difference is, why van Inwagen revises the traditional Fregean interpretation of the claim that the affirmation of existence is a denial of the number 0, and what the consequences of this revision are, I will begin by offering a very brief sketch of the Fregean view.

As we will see in more detail in the next chapter, Frege, Russell, and Quine are usually associated with the view that the existence of an object X does not consist in X's having a property but in a property Y's having the property of being instantiated. Now, one of Frege's classical arguments to support this view is that statements of number are statements about properties and that existential statements are statements about properties as well, because statements of existence are statements of number.[18]

The argument runs like this. According to Frege, a statement such as 'Jane Doe has three siblings' is logically equivalent to the statement of number 'The number of Jane Doe's siblings is three'.[19] If this is the case, the question is what the bearer of the property of being three is. In other words, what is the bearer of the property which is expressed by the predicate 'The number of siblings... is three'?[20] Neither Jane Doe nor one of her siblings has the property of being three. Frege's answer to this question is that statements of number are not statements about objects but statements about properties. The bearer of the property of being three is the property of being Jane Doe's sibling. Thus, a statement of number such as 'The number of Jane Doe's siblings is three' attributes the property of being exemplified or instantiated three times to the property of being Jane Doe's sibling.[21] Expressed in general terms: A statement of number attributes the property of being exemplified or instantiated n-times to the property of being F.

Now, according to Frege, existential statements as well are statements of number and must be analyzed in an analogous way. Thus, an existential statement such as 'Jane Doe's sibling exists' is logically equivalent to the statement

---

17 Cf. ibid., 483.

18 Cf. Johannes Hübner, *Einführung in die theoretische Philosophie* (Stuttgart: Verlag J.B. Metzler, 2015), 186–187. In what follows, I am indebted to the very helpful explanation of Frege's position provided by ibid., 187–188.

19 Cf. ibid., 187.

20 Cf. ibid.

21 On this see also section 3 of Edward N. Zalta, "Frege's Theorem and Foundations of Arithmetic," ed. Edward N. Zalta, Winter 2000 ed., *The Stanford Encyclopedia of Philosophy* (2020), https://plato.stanford.edu/archives/win2020/entries/frege-theorem/.

of number 'The number of Jane Doe's siblings is greater than 0'.[22] The statement of number 'The number of Jane Doe's siblings is greater than 0' attributes the property of being exemplified or instantiated at least once to the property of being Jane Doe's sibling. In the same way, a negative existential statement such as 'Unicorns do not exist' is logically equivalent to the statement of number 'The number of unicorns is 0' which states that the times the property of being a unicorn is instantiated or exemplified is 0. Thus, existential statements are statements of number, and as such they are statements about properties and not objects. Consequently, according to the Fregean doctrine 'Affirmation of existence is a denial of the number 0', a number is not attributed to objects but to a property. The number indicates how many times a property is instantiated or exemplified.

We are now in the position to see why van Inwagen is forced to revise the standard interpretation of the Fregean doctrine that the affirmation of existence is a denial of the number 0. With the traditional Fregean interpretation of this doctrine, premise (2) entails that 'exists' is not predicated of objects but of properties.[23] Thus, when one speaks, for example, about the existence of horses, one does not—despite appearances—speak about horses but about the property of being a horse.[24] One does not attribute a number to objects but to a property.

But if this is the case, van Inwagen's argument for the univocity of 'exists' does not work. It does not work because if premise (2) is interpreted in this way, it does not follow that 'exists' does not have different meanings when it is applied to *objects* in different logical or ontological or metaphysical categories. This conclusion does not follow because 'exists' is not applied to objects but to *properties*. Thus, van Inwagen has to revise the traditional interpretation of the Fregean claim that the affirmation of existence is a denial of the number 0 by giving up the idea that 'exists' is a second-order or second-level predicate, i.e., a predicate of properties rather than objects.[25] For example, if one says that 'The number of horses is 1 or more', one indeed predicates of horses that they number more than 0.[26] Only on the basis of this interpretation of premise (2), are existential statements such as 'Unicorns do not exist' or 'Horses exist' logically equivalent to statements of number such as 'The number of unicorns is 0' or 'The number of horses is 1 or more'. Whereby the conclusion follows that 'exists' does not have different meanings when it is applied to objects in dif-

22 Cf. Hübner, *Einführung in die theoretische Philosophie*, 188.
23 Cf. van Inwagen, *Metaphysics*, 299.
24 Cf. ibid.
25 Cf. van Inwagen, "Being, Existence, and Ontological Commitment," 484.
26 Cf. ibid., 483; van Inwagen, *Metaphysics*, 299.

ferent logical or ontological or metaphysical categories. Thus, to rebut an important objection against his argument for the univocity of 'exists', van Inwagen concedes that 'exists' is predicated of objects and that being or existence is a property of objects.[27]

Now, this brings us full circle to the problem that seemingly any attempt to identify and reduce the existence of an object to that object's having a property is vulnerable to counterexamples of objects which lack the relevant property and yet exist. Van Inwagen believes that he can solve this problem through the identification and reduction of existence with and to an object's property of being identical with itself. His reasoning goes like this: Given his interpretation of the Fregean doctrine that affirmation of existence is a denial of the number 0, the existence of an object can be identified with and reduced to that object's having the property 'that the number of things that are it is not 0'.[28] Now, given that being and existence are the same and that 'exists' is univocal in van Inwagen's view, this proposal is vulnerable to the objection that it does not successfully reduce the existence of an object to that object's having a property, because the relevant property is an object's property whereby the number of things that *are* it is not 0.[29] So, it seems that reductionism is unsuccessful because it does not eliminate a reference to being or existence.

To dispel this worry, van Inwagen suggests that the existence of an object can be identified with and reduced to that object's having the property 'that the number of things that are numerically identical with it is not 0'.[30] Now, since the number of things that are numerically identical with an object cannot be two or any greater number, the property 'that the number of things that are numerically identical with it is not 0' can be simplified to the property 'that the number of things that are numerically identical with it is one' and since the only thing that something can be identical with is itself, the property can be simplified further to the property 'that it is identical with itself'.[31] Thus, the existence of an object can be identified with and reduced to that object's having the property of being identical with itself. [32] This reductive view is not vulnerable

27 Cf. "Being, Existence, and Ontological Commitment," 483–484; *Metaphysics*, 299–301.
28 Cf. *Metaphysics*, 300.
29 For this reason one should distinguish van Inwagen's view from the view that an object's existence consists in that object's having the property of being identical with something that is or exists which has been proposed by Saul A. Kripke in Kripke, *Reference and Existence*.
30 Cf. van Inwagen, *Metaphysics*, 300.
31 Cf. ibid., 300–301.
32 Cf. ibid., 301.

to counterexamples because there are no objects which lack the property of identity and yet exist.

To sum up: Van Inwagen's view that the existence of an object consists in that object's having the property of being identical with itself is motivated by his belief that it can solve two problems. First, his view can solve the problem that his 'univocity of number' argument works only if that which accounts for the existence of objects is a property of objects and not of properties. Van Inwagen considers this argument to be important to undermine the idea that we use 'exists' in different senses in our ordinary ways of talking, which in turn motivates an ontological pluralism according to which there are different modes of being or existence. Any version of ontological pluralism, i.e., any version of the view that objects exist in different ways, challenges van Inwagen's view on existence according to which all there is to say about an object's existence is adequately expressed by the existential quantifier of formal logic. If it is true that objects exist in different ways, van Inwagen's view is false because there is more to say about an object's existence than that it exists. A pure quantificational account of existence cannot capture the fact that objects exist in different ways. Thus, by conceding that an object's existence consists in that object's having a property, namely, the property of being identical with itself, van Inwagen can rescue the 'univocity of number' argument which plays an important part in his argumentative strategy against ontological pluralism.

Second, by stating that an object's existence consists in that object's having the property of being identical with itself, van Inwagen can present a version of the general view that the existence of an object consists in that object's having a property which is not vulnerable to possible counterexamples. There are no objects which lack the property of being identical with themselves and yet exist. Therefore, a second problem is solved.

2.2 Three Problems

2.2.1 Reductionism Does Not Work

The first problem with van Inwagen's view is that its reductive explanatory strategy fails by its own standards for a successful reduction. Following Quine's 'method of paraphrase', van Inwagen argues that the identification and reduction of an object's existence with and to that object's having the property of being identical with itself is successful, if any sentence in which 'exists' is applied to an object can be replaced with a sentence in which no such term occurs.[33] In particular, van Inwagen suggests that it should be possible to replace every sentence in which 'exists' is applied to an object with a sentence in which 'is identical with itself' is applied to an object.[34] Furthermore, an important criterion for a successful reduction to which he commits himself is that the essential meaning of a sentence is preserved in this process of paraphrasing, rewriting, or translating it.[35]

Now, by using some of van Inwagen's own examples, I will show that his reductive explanatory strategy fails by his own standards for a successful reduction because the essential meaning of statements which apply 'exists' to an object is not preserved by statements in which 'is identical with itself' is applied to an object. In other words, van Inwagen must show that all sentences which use 'exists' can be paraphrased or translated into quantificational sentences which do not have the term 'exists' without changing the meaning of the original sentence which applies the term 'exists' to an object.[36] I will argue that this paraphrasing strategy does not work for all cases. It is not possible to paraphrase or translate every sentence which uses 'exists' into a sentence which uses the existential quantifier and a first-order identity predicate, i. e., a sentence of the logical form $\exists x$ ($x = \ldots$) without changing the meaning of the whole sentence.[37]

I will concentrate on three examples which van Inwagen himself proposes to show how his paraphrasing strategy works:[38]

(1) 'Dragons do not exist' means 'Everything is not (identical with) a dragon'
(2) 'I think, therefore I am' means 'I think, therefore not everything is not I' or 'I think, therefore I am myself'
(3) 'It is a great mystery why there should be anything at all' means 'It is a great mystery why it is not the case that everything is not (identical with) anything'

33 Cf. van Inwagen, "Being, Existence, and Ontological Commitment," 478–479; *Metaphysics*, 314–319. For a similar view on what a successful reduction requires, see also McDaniel, *The Fragmentation of Being*, 224.
34 Cf. van Inwagen, "Being, Existence, and Ontological Commitment," 478–479.
35 Cf. *Metaphysics*, 314.
36 Cf. "Being, Existence, and Ontological Commitment," 478–479.
37 Cf. Hübner, *Einführung in die theoretische Philosophie*, 188–189; Miller, *The Fullness of Being*, 7–9.
38 Cf. van Inwagen, "Being, Existence, and Ontological Commitment," 478–479; *Metaphysics*, 301.

Let us begin with (1). There is a widespread consensus that dragons do not exist. There are no dragons in actuality. Therefore it is not surprising to discover that van Inwagen agrees that it is true that dragons do not exist.[39] What about the sentence 'Everything is not (identical with) a dragon'? If the sentence 'Dragons do not exist' is true, the sentence 'Everything is not (identical with) a dragon' must be true as well, otherwise it would have a different truth value than the sentence 'Dragons do not exist'. This would be fatal for van Inwagen's view as it would show that the paraphrasing strategy it employs does not work in this case because the essential meaning of the sentence 'Dragons do not exist' is not preserved by the sentence 'Everything is not (identical with) a dragon'.

Now, it seems that van Inwagen must deny that the sentence 'Everything is not (identical with) a dragon' is true. Why? Consider the dragons Fafnir and Elliot. Fafnir *is* a dragon according to Nordic mythologies and Elliot *is* a dragon according to Disney's *Pete's Dragon*.[40] Therefore, while the claim that dragons do not exist is true, the claim that everything is not (identical with) a dragon is false. The number of things which are Fafnir is not 0 but 1 and the number of things which are Elliot is not 0 but 1. From van Inwagen's interpretation of the Fregean doctrine that affirmation of existence is a denial of the number 0, it follows that there are at last two dragons which exist, because both Fafnir and Elliot have the property 'that the number of things that are it is not 0'. Both Fafnir and Elliot are objects which are identical with themselves. Consequently, van Inwagen's paraphrasing strategy fails. The sentence 'Dragons do not exist' is true. However, if this sentence is paraphrased into 'Everything is not (identical with) a dragon' its meaning is not preserved because it has a different truth value.

Let us have a look at van Inwagen's second example (2). The problem with this example is that the sentence 'I think, therefore I am' means something different than the sentences 'I think, therefore not everything is not I' or 'I think, therefore I am myself'. To see the difference, let us consider what Descartes wants to demonstrate with the proposition 'I think, therefore I am'. Descartes intends to prove that one cannot question or doubt one's own existence. Descartes' proof works this way: When I am thinking about the possibility of being deceived about my own existing, it is me that is thinking. Now, I cannot doubt the fact that I am thinking when I am thinking. Furthermore, the fact that I am thinking presupposes that I exist. I cannot think if I do not exist. Thus, I can infer from the

39 Cf. "Being, Existence, and Ontological Commitment," 478–479; *Metaphysics*, 125.

40 For a similar objection, see "Being, Existence, and Ontological Commitment," 478 n.12. Van Inwagen proposes a possible solution to this problem with which I will deal below (see 2.2.3).

fact that I am thinking that I exist. And since I cannot doubt the fact that I am thinking, I cannot doubt the fact that I exist. So, Descartes's basic rationalistic idea is that the exercise of an activity like thinking can prove that we exist and that this proof is better suited to rebut skeptical worries than arguments which use empirical premises, because this proof does not rely on the exercise of sensory activities, like seeing, about which we could be deceived.[41]

Now, let us substitute the sentence 'I think, therefore I am' with the sentence 'I think, therefore not everything is not I' and replace forms of 'existence' with forms of 'to be identical with itself' within Descartes's proof. Descartes's paraphrased proof runs like this: When I am thinking about the possibility of being deceived about my being identical with myself, I am thinking. Now, I cannot doubt the fact that I am thinking when I am thinking. However, the fact that I am thinking presupposes that not everything is not I. I cannot think if everything is not I. Thus, I can infer from the fact that I am thinking that not everything is not I. And since I cannot doubt the fact that I am thinking, I cannot doubt the fact that not everything is not I.

Whatever we think about the soundness or merit of Descartes's original argument or proof and the paraphrased proof, it should be obvious that the paraphrased version proves something different, if anything at all. At best, the paraphrased version proves that I cannot doubt consistently that not everything is not I. Or, to put it simpler, what is proved is that I cannot doubt my being identical with myself. Formulated in this way, this proof looks more like a psychological exercise for someone who is troubled about their own identity. What is missing in this paraphrased version is an argumentative step which aims to prove one's existence.

In the original proof, Descartes's argument is that the fact that I am *thinking* presupposes that I *exist*. What is made explicit is that the exercise of thinking presupposes having existence. In the paraphrased proof, it is misleading to state that the fact that I am thinking *presupposes* that not everything is not I. Rather, I can infer from the fact that I am thinking that not everything is not I. But again, what is proved thereby is at best that I am identical with myself and not that I exist. So, strictly speaking, this argument only demonstrates why I cannot consistently doubt my being identical with myself but not why I cannot consistently doubt that I exist. This shows that the meaning of the sentence 'It think, therefore I am' is not preserved by the paraphrase 'I think, therefore not everything is not I'.

---

41 Descartes, of course, also considers the possibility that we could be deceived about this as well, but this complication does not matter for the point I want to make here.

This brings us to the third example (3). Let us imagine a situation where it is plausible that we affirm the proposition ‘It is a great mystery why there should be anything at all’. Maybe we are looking at the stars on a warm summer night and this sight makes us aware of the miracle of the mere presence of things, i.e., of the wonder *that* there are things. We begin to recognize that we should not simply take it for granted that there are actual things, i.e., that things exist. And to express this feeling of wonder and awe about the mere presence or actuality of things, we utter a sentence like ‘It is a great mystery why there should be anything at all’.

Maybe a similar occasion is the birth of a child. By uttering the proposition ‘It is a great mystery why there should be anything at all’ we do not give expression to our wonder about how this child was generated. Usually there is a plausible story about that. No, with awe and wonder we are just fascinated by the mere fact that the baby is, i.e., that there is a newborn life. The birth of the baby makes us aware of something we usually take for granted, namely, *that* we are, i.e., that we *exist*. We become aware of the mere presence or actuality of living things.

If van Inwagen is right about the idea that it is possible to replace every sentence in which ‘exists’ occurs with a sentence in which ‘is identical with itself’ occurs without altering the meaning of the original sentence, it should be possible to express the same kind of awe and wonder with the sentence ‘It is a great mystery why it is not the case that everything is not (identical with) anything’. I think there is a strong intuition that this does not work. It does not work because we express awe and wonder about a different thing with the latter sentence.

Imagine you witness someone looking at the stars or at a newborn baby, and they utter the sentence ‘It is a great mystery why it is not the case that everything is not (identical with) anything’. Besides the fact that you might think that this is a rather peculiar way to express awe and wonder, what would you think this person is astonished about? I believe it is quite plausible to assume that this person is struck by the sheer plurality of things. It seems that this person realizes in such a moment something which they had taken for granted, namely, that there is something which diversifies and individuates things. This star is different from that star. The baby is different from their parents, their diapers, the bed etc. and their mother is different from their father, and both are different from the nurse etc.

Whatever the exact phenomenological description of this feeling of awe and wonder is, I think there is a strong intuition in favor of the view that it is different from the feeling of awe and wonder which we express with the sentence ‘It is a great mystery why there should be anything at all’. In the latter case, the awe and wonder are about the *thatness* of things, i.e., the mere presence or actuality of things. In the case of the sentence 'It is a great mystery why it is not the case that everything is not (identical with) anything' the awe and wonder is about the *whatness* of things, i.e., about the fact that there is a plurality of things which differ from each other. For this reason, I think that van Inwagen is wrong. The sentences do not mean the same thing. They give expression to different kinds of awe and wonder.

To sum up: These three examples indicate that van Inwagen's reductive explanatory strategy fails by his own standards for a successful reduction. An object's existence cannot be identified with and reduced to that object's having the property of being identical with itself. Existence cannot be identified with and reduced to self-identity.

2.2.2 Existence Is Not an Essential Property of an Object

In the previous section, I have argued that van Inwagen's attempt to identify and reduce an object's existence with and to that object's having the property of being identical with itself fails by his own standards for a successful reduction. Now, let us assume for the sake of argument that the reduction *is* successful. In this case, van Inwagen's view on what it is for an object that exists to exist is confronted with an even more devastating second problem.

The second problem with van Inwagen's version of the reductive-object view is the following: If for an object to exist is to have the property of being identical with itself, it follows that every object which has the property of being identical with itself exists necessarily, because self-identity is an essential property of objects. But objects which have existence and therefore exist do not have existence necessarily and therefore exist necessarily. Objects which have existence could possibly not have existence and not exist.

How could van Inwagen try to rebut this objection? Well, he could try to challenge the claim that self-identity is an essential property of an object that has it. Thus, he could try to argue that having the property of being identical with itself is accidental to an object that is identical with itself. This attempt doesn't look very promising, however, as becomes evident when we consider how contemporary philosophers typically characterize the distinction between essential and accidental properties (see 1.2.2):

Basic modal characterization (BMC):

> *P* is an essential property of an object *o* just in case it is necessary that *o* has *P*, whereas *P* is an accidental property of an object *o* just in case *o* has *P* but it is possible that *o* lacks *P*.

Possible world characterization (PWC):

> *P* is an essential property of an object *o* just in case *o* has *P* in all possible worlds, whereas *P* is an accidental property of an object *o* just in case *o* has *P* but there is a possible world in which *o* lacks *P*.

Existence-conditioned modal characterization (ECMC):

> *P* is an essential property of an object *o* just in case it is necessary that *o* has *P* if *o* exists, whereas *P* is an accidental property of an object *o* just in case *o* has *P* but it is possible that *o* lacks *P* and yet exists.

Modified possible world characterization (MPWC):

> *P* is an essential property of an object *o* just in case *o* has *P* in all possible worlds in which *o* exists, whereas *P* is an accidental property of an object *o* just in case *o* has *P* but there is a possible world in which *o* lacks *P* and yet exists.

According to the BMC, self-identity is an accidental property of an object just in case that this object has the property to be identical with itself, but it is possible that it could lack this property. Yet, it is not possible for an object which has the property to be identical with itself not to have it. It makes no sense to say, for example, that Angela Merkel who has the property to be identical with herself could also not have this property and consequently not be identical with herself.

We get the same result if the PWC, the ECMC, or the MPWC are used to describe the distinction between essential and accidental properties. According to the PWC, the property of being identical with itself is not an accidental property of an object that has it because there is no possible world in which an object which has the property to be identical with itself lacks this property. According to the ECMC, the property of being identical with itself is not an accidental property of an object that has it, because it is not the case that it is possible that an object which has the property to be identical with itself lacks this property and yet exists. And according to the MPWC, the property of being identical with itself is not an accidental property of an object that has it because it is not the case that there is a possible world in which the object lacks the property and yet exists. To sum up: The attempt to challenge the claim that self-identity is an essential property of an object that has it by arguing that it is an accidental property of an object that has it is doomed to fail.

Consequently, if the existence of an object consists in that object's having the property of being identical with itself and this property is not an accidental property of it, it is an essential property of it. Now, according to the considered standard modal ways of characterizing the distinction between essential and accidental properties, this means that an object which has the property of being identical with itself has this property necessarily. An object which has the property of being identical with itself cannot fail to have this property. Thus, an object which is identical with itself is necessarily identical with itself.

Now we can see the problem: Objects which have existence and therefore exist do not have existence necessarily and therefore exist necessarily. Objects which have existence could possibly not have existence and not exist. Consequently, the existence of an object X does not consist in X's having the property of being identical with itself because X has the property of being identical with itself necessarily, but it does not have existence necessarily. To suppose that objects which have existence have existence necessarily is absurd. Such a claim is clearly belied by our experience. Our experience tells us that the existence of objects is contingent. Objects which have existence and exist exist contingently. It is possible for them not to have existence and not to exist. For example, the fact that the apple which had existence before my lunch does not exist anymore after being eaten by me at lunch shows that the apple did not have existence necessarily but contingently. From the very fact that the apple is no more, we can infer that it was possible for the apple which had existence not to have had existence.

This simple fact of an apple's ceasing to exist cannot be explained by the view that the existence of an object consists in that object's having the property of being identical with itself. The explanation would be that the apple ceases to exist in virtue of losing the property of being identical with itself. But since the property of being identical with itself is an essential property of an object, this property cannot be lost. It was not possible for the apple which was identical with itself not to be identical with itself.

The upshot of this is that van Inwagen's view fails to be convincing because it implies that objects which have existence exist necessarily. Experience tells us, however, that this is not the case. Objects which have existence could possibly not have existence and not exist. The attempt to identify and reduce the existence of an object with and to that object's having the property of being identical with itself fails because having self-identity is essential to an object but having existence is not.

2.2.3 Not All Objects Which Are Identical with Themselves Exist

A third problem with van Inwagen's view is that it is vulnerable to counterexamples. Not all objects which are identical with themselves exist. For example, fictional objects like Sherlock Holmes have the property of being identical with themselves but do not exist.[42] Such counterexamples contradict the claim that the existence of an object consists in that object's having the property of being identical with itself.

This third problem is generated by van Inwagen's attempt to solve an important problem with all variants of the general view that the existence of an object consists in that object's having a property (other than existence). As I have explained above (see 2.1.1), the problem with any such view is that it is vulnerable to counterexamples. Whatever property of an object the existence of an object is identified with and reduced to, it seems that there are objects which lack this property and yet exist. Van Inwagen's view seems to have a clever solution for this problem. It is not vulnerable to counterexamples of this kind because every object that exists has the property of being numerically identical with itself. It would be absurd to claim that an object lacks the property of being identical with itself and yet exists.

It is true that there are no objects which lack the property of being identical with themselves and yet exist. Nevertheless, van Inwagen's view is vulnerable to a different kind of counterexamples. There are objects which have the property of being identical with themselves and yet *do not* exist. In other words, not all objects which are identical with themselves exist. Consequently, the existence of an object does not consist in that object's having the property of being identical with itself because there are objects which are identical with themselves and yet do not exist.

What objects have the property of being identical with themselves and yet do not exist? Plausible candidates seem to be mythological or fictional objects like Pegasus, Fafnir, Elliot, or Sherlock Holmes.[43] From what we have seen so far, it is clear that van Inwagen is committed to the claim that a fictional object like Sherlock Holmes has the property of being identical with itself.[44] Given his interpretation of the Fregean doctrine that affirmation of existence is a denial of the number 0 (see 2.1.3), the property 'that the number of things that are it is not 0' is a property of an object which can be simplified to the property 'that the

42 To be more precise, fictional objects like Sherlock Holmes do not exist in the sense of being actual. Sherlock Holmes is not an actual human being or detective. Nevertheless, Sherlock Holmes exists in the sense that he exists as a fictional character, i.e., as a being in the mind. I will deal with this issue in more detail below. I am thankful to Jeremy Skrzypek for making me aware of the need to address this worry.

43 In what follows, I use the terms 'fictional' and 'mythological' interchangeably. If there is a difference between mythological and fictional objects, it does not matter for my argument.

44 Cf. van Inwagen, *Metaphysics*, 309.

number of things that are numerically identical with it is not 0' which in turn can be simplified to the property 'that the number of things that are numerically identical with it is one'.[45] And since the only thing that anything can be identical with is itself, the property can be further simplified to the property of an object 'that it is identical with itself'.[46] Now, since van Inwagen concedes that the number of things that are Sherlock Holmes is not 0 but 1, he is committed to the claim that Sherlock Holmes is an object which has the property of being identical with itself.[47] But he also acknowledges that Sherlock Holmes does not exist because there is no person or human being which could be correctly labeled 'Sherlock Holmes, the famous detective'.[48] Consequently, it seems that van Inwagen must accept that there are objects which are identical with themselves but do not exist, namely, fictional objects like Sherlock Holmes.

To avoid this conclusion, van Inwagen argues that the proposition 'Sherlock Holmes does not exist' is ambiguous. It can either mean 'There is no such person, or detective, or man, or Londoner as Holmes', or 'There is no such fictional character as Holmes'.[49] Thus, according to van Inwagen, the proposition that there is no person as Sherlock Holmes is true because the number of persons that are identical with Sherlock Holmes is 0, and the proposition that there is no fictional character as Sherlock Holmes is false because the number of fictional characters which are identical with Sherlock Holmes is not 0 but 1.[50] Consequently, there is no object which has the property of being identical with itself but yet does not exist. There is no person or human being labeled 'Sherlock Holmes, the famous detective' who has the property of being identical with themselves. A person who has this property does not exist. It is true that there is a fictional object named 'Sherlock Holmes' which has the property of being identical with itself. However, this fictional object is not a non-existing object, but exists.[51] Thus, van Inwagen argues, fictional objects like Sherlock Holmes are no counterexamples. Fictional objects are objects which have the property of being identical with themselves and are objects which exist.

But this rebuttal is hardly convincing because a proposition such as 'The fictional object Sherlock Holmes exists' is ambiguous as well. It can either mean 'The fictional character Sherlock Holmes exists in the sense of being actual,

45 Cf. ibid., 300–301.
46 Cf. ibid., 301.
47 Cf. ibid., 309.
48 Cf. ibid.
49 Cf. ibid.
50 Cf. ibid.
51 Cf. ibid.

i.e., its existence is not dependent on the activities of a human mind', or it can mean 'The fictional character Sherlock Holmes exists in the sense of existing in the human mind, i.e., its existence is dependent on the activities of a human mind'.

Consequently, a critic of van Inwagen's view can simply reformulate their original objection in the following way: Fictional objects like Sherlock Holmes are objects which have the property of being identical with themselves and yet do not exist in the sense of being actual. Sherlock Holmes is not an actual human being whose existence does not depend on the activities of a human mind. Rather, he is a fictional human being or character whose existence does depend on the activities of a human mind. Consequently, there are objects which have the property of being identical with themselves and yet do not exist in the sense of being actual. It follows that the existence of an object cannot be identified with and reduced to that object's having the property of being identical with itself, because there are fictional objects like Sherlock Holmes which have the property of being identical with themselves and yet do not exist in the sense of being actual.

To rebut this reformulated objection, van Inwagen must reject the idea that actual and fictional objects exist in different senses or ways. Instead, he must argue that everything that exists exists in the same sense. And this is in fact his position:

> That is to say, there does *exist* such a fictional character as Sherlock Holmes. He is as much a part of the World as is any of the short stories and novels in which he 'occurs,' and *they* are as much a part of the World as their author and the pen with which he wrote them.[52]

According to van Inwagen, everything that exists exists in the same sense because 'exists' is univocal and being is the same as existence.[53] Consequently, there is no such thing as different modes of being or existence. As we have seen above (see 2.1.3), van Inwagen is forced to reject the ideas that 'exists' is used in many senses and that objects exist in different ways because these ideas give expression to something which cannot be captured by the quantificational account of existence he wants to defend.[54] It makes no sense to say that the term 'that the number of things which are identical with it is not 0' is used in different senses when it is applied to objects in different logical or onto-

52 Ibid. (Italics in the original).
53 Cf. van Inwagen, "Being, Existence, and Ontological Commitment," 480–492.
54 Cf. ibid., 492–499.

logical or metaphysical categories.[55] Likewise it makes no sense to say that objects have the property that the number of things which are identical with them is not 0 in different ways or that objects are numerically identical with themselves in different senses.

If everything that exists exists in the same sense, the question is of course in which sense does everything that exists exist? Van Inwagen must reject the view that everything that exists exists in the sense of being actual because he wants to affirm that fictional objects like Sherlock Holmes exist, and it is clear that fictional objects do not exist in the sense of being actual. They are called 'fictional' objects precisely because they are not actual objects. Being a fictional object implies not being an actual object. 'Fictional' qualifies and signals that the object it is applied to is not actual. So, if everything that exists exists in the same sense and if fictional objects like Sherlock Holmes exist, everything that exists must exist in a sense which also covers fictional objects like Sherlock Holmes. Consequently, van Inwagen's suggestion is that everything that exists exists in the same sense, namely, in the sense that the number of things that are it is not 0. Everything that exists exists in the sense of being identical with itself.

But if everything that exists exists in the same sense and this is the one and only sense in which everything that exists exists, how does van Inwagen explain the difference between actual and fictional objects? For example, how does he explain the difference between an actual human being like Angela Merkel and a fictional character or human being like Sherlock Holmes? According to van Inwagen's view on what it is to exist, Angela Merkel and Sherlock Holmes both exist. And not only do they both exist, but they also exist in the same sense, namely, in the sense that the number of things which are identical with them is not 0 but 1. Yet there must be a difference because if we were to count all the human beings or persons that exist in the universe, a human being or person who could be correctly labeled 'Sherlock Holmes, the famous detective' would not be among them.[56]

Now, since everything that exists exists in the same sense and since there are no modes of existence or being, the explanation for the difference cannot be that Angela Merkel exists mind-independently, i.e., that her existence is not dependent on an activity of the human mind, and that Sherlock Holmes exists mind-dependently, i.e., that his existence is dependent on an activity of the human mind. According to van Inwagen, if one explains the difference in this way, one commits the fallacy of ascribing to the being or existence of a thing what properly

---

55 Cf. ibid., 482.
56 Cf. van Inwagen, *Metaphysics*, 309.

belongs to its nature.[57] Consequently, if there is a difference between an actual human being or person like Angela Merkel and a fictional human being or character like Sherlock Holmes, this difference must be accounted for in terms of a difference in their respective natures.

To sum up: According to van Inwagen, the difference between actual and fictional objects does not consist in having different sorts or kinds of being or existence—existence in the sense of actuality vs. existence in the sense of existing in the mind—but rather in having different sorts or kinds of nature.[58] Like van Inwagen's explanation for what it is for an object to exist, this explanation is reductive in character: Whatever is supposed to be explained with the help of different modes or ways of existing can be explained with the help of differences between the natures of the objects that exist. Talk about different ways in which objects exist can be paraphrased to talk about differences between the natures of the objects that exist.

However, this explanation for the difference between an actual object like Angela Merkel and a fictional object like Sherlock Holmes does not work. It does not work because there is no difference between the natures of Merkel and Holmes. The essence or nature of both is humanity. If you ask, 'What is it?', the correct answer in both cases is 'a human being'. Now, to argue that their essences or natures are different because 'actual' is part of Merkel's nature but not of Holmes's and 'fictional' is part of Holmes's nature but not of Merkel's, does not work either. It does not work because it implies that Angela Merkel is actual in virtue of her nature, i.e., that it is part of her nature to be actual. In other words, the explanation offered by van Inwagen implies that it is part of the essence or nature of an actual object to be actual. It follows that, in van Inwagen's view, Angela Merkel like any other actual object is necessarily actual, i.e., cannot fail to have actuality. But such a position is absurd, as I have shown in detail in the previous section, because it conflicts with our experience that objects which exist and are actual can cease to have existence and cease to be actual objects.[59]

---

57 Cf. "Being, Existence, and Ontological Commitment," 477 n.10, 481 n.16.

58 Cf. ibid., 477.

59 Timothy Pawl suggested to me that maybe there is an alternative route for van Inwagen to argue that there is a difference between the natures of Merkel and Holmes. According to Pawl, van Inwagen could say that Merkel is a concrete, flesh-and-blood human. She falls under the kind 'living organism'. In contrast, Sherlock is an abstract object, not flesh-and-blood. He does not fall under the kind 'living organism'. *In the fiction*, he is a living organism like Merkel. *But in actuality*, the real Sherlock is an abstract entity. Van Inwagen could further argue that he distinguishes between a fictional entity's *having* a property and *holding* a property.

The upshot of this is that van Inwagen's attempt to defend the claim that everything that exists exists in the same sense fails because he cannot explain the difference between actual and fictional objects in terms of a difference between their essences or natures. Consequently, he has not successfully rebutted the reformulated version of the objection that there are objects which have the property of being identical with themselves and yet do not exist in the sense of being actual, namely, fictional objects like Sherlock Holmes. It follows that the existence of an object cannot be identified with and reduced to that object's having the property of being identical with itself because there are fictional objects like Sherlock Holmes which have the property of being identical with themselves and yet do not exist in the sense of being actual.[60]

Thus, Sherlock *has* the property of being a fictional entity, but Sherlock only *holds* the properties that he has in the fiction (e.g., a concrete particular made out of flesh and blood). So, while he isn't really a detective—he doesn't *have* the property—he does stand in a different relation to the property of being a detective, the relation van Inwagen dubs 'holding'. My response is that this way to claim that Merkel and Holmes have different natures is not available for van Inwagen in this dialectical context because it conflicts with his rejection of ontological pluralism and his commitment to the metaontological claims that everything that exists exists in the same sense; that is, 'exists' is univocal and being and existence are the same. To say that '*In the fiction*, Holmes is a living organism, but *in actuality*, Holmes is an abstract entity' is just another way to say that Holmes exists in two different ways; namely, he exists in the sense of being fictional *and* in the sense of being actual. Thus, whether or not this move would help van Inwagen to justify his claim that Holmes and Merkel have different natures—I have my doubts here as well, but I will not pursue the issue further—, would require van Inwagen to admit the falsity of some of his metaontological claims.

60 Jeremy Skrzypek made me aware that van Inwagen's view is also vulnerable to an objection similar to Bradley's Regress which I wield against the property view in the next chapter (see 3.2.2): An infinite regress is triggered because if to exist is to have the property of being self-identical, it follows that the property of being self-identical must exist, which in turn must have the property of being self-identical to exist, which in turn must have the property of being self-identical to exist and so on *ad infinitum*.

3 The Property View

The remaining three views I discuss in this first part of my book also employ a reductionist explanatory strategy. What distinguishes them from the reductive-object view is that they reject the idea that the existence of an object X can be identified with and reduced to X's having the property of being identical with itself or any other property which X has. Rather, what unites these views is the idea that the existence of an object is identical with and can be reduced to a property which *something else Y* has. In other words, these views maintain that the existence of an object consists in something else Y's having a property.[1] In this and the following two chapters, I reconstruct and criticize three different versions of this view.

3.1 The Core Idea of the View

In this chapter, I discuss a version of the view that the existence of an object X consists in something else Y's having a property with which contemporary analytic philosophers are probably most familiar with. According to this view, the existence of an object X consists in a property Y's having the property of being instantiated. Such a view is standardly associated with a tradition of thinkers including Frege, Russell, and Quine. Let us call this view the 'property' view.

The core idea of the property view can be summarized as follows:[2] Let us grant that Angela Merkel is an object that exists. What is it for Angela Merkel to exist? The answer of the property view to the nature question is that the ex-

1 James Dominic Rooney has suggested to me that the core idea of the property view could also be formulated in terms of the distinction between an object's intrinsic and extrinsic properties. If one makes use of this terminology, the core idea of the property view can be expressed as follows: An object X exists in virtue of having existence as an extrinsic property. I think this is an intriguing proposal. But this way to formulate the core idea requires clarifying what an extrinsic property exactly is in contrast to an intrinsic property. And as far as I can see, such an explanation would end up with something very similar with what I am saying, e.g, that existence is an extrinsic property of X because X has existence in virtue of the way something else Y is, namely, that Y is such that it contains X. So, from my point of view, not much is gained by this proposal and the need to explain and to defend the idea of extrinsic properties would be more distracting than helpful.
2 I am loosely following van Inwagen's presentation of the view, see van Inwagen, *Metaphysics*, 296–299.

https://doi.org/10.1515/9783110979879-006

istence of Angela Merkel consists in a property's having the property of being instantiated.

The question is, of course, what kind of property is such that the existence of Angela Merkel can be identified with and reduced to that property's having the property of being instantiated? Well, it must be a property which is such that the instantiation of this property is sufficient for Angela Merkel and only for Angela Merkel to exist. In other words, it must be a property which individuates Angela Merkel, i. e., a property of which Angela Merkel is the only possible instance. For example, in the case of Angela Merkel it could be argued that the relevant property is the property of being a woman named 'Angela Merkel' who was born on the 17th of July 1954 in Hamburg and who was the 8th chancellor of Germany after the Second World War.[3] Thus, the existence of Angela Merkel consists in this property's having the property of being instantiated.

Before I proceed, two worries need to be addressed. A first worry is semantic in nature. It could be objected that, according to my presentation, the property view presupposes something like a Russellian kind of descriptivism according to which a singular existential which involves a proper name—e. g., 'Angela Merkel exists'—can be paraphrased to a statement which involves a definite description (e. g., being a woman called 'Angela Merkel' who was born on the 17th of July 1954 in Hamburg and who was the 8th chancellor of Germany after the Second World War). But it seems that such a descriptivism is vulnerable to objections which, for example, have been advanced by Saul A. Kripke.[4]

To this objection I would like to respond that the issue of a descriptive analysis of proper names is a complex semantic discussion which is beyond the scope of this metaphysical inquiry. So, for the sake of argument, I simply assume that proponents of the property view have a way to deal with this or similar semantic objections or that the property view can be proposed without a commitment to a descriptive analysis of proper names. Since nothing of my criticism of the property view hinges on these semantic issues, I leave them aside.

A second worry is metaphysical in nature. It could be objected that I have presented the property view in such a way that it generates the metaphysical problem of individuation. If the property, whose having the property of being instantiated is supposed to explain the existence of an object, must be such that its only possible instance is the object whose existence should be explained, this view requires an account of individuation. Thus, the property view must be

3 Cf. McGinn, *Logical Properties*, 17–18; van Inwagen, *Metaphysics*, 297. I follow van Inwagen's suggestion in treating this property as a single 'compound' property, and not as a class of properties which comprises a plurality of properties as its members.
4 Cf. Saul A. Kripke, *Naming and Necessity* (Oxford: Blackwell, 1980).

able to explain how a property could be such that it individuates an object, i.e., that it could be such that it has only one object as a possible instance.

I respond to this worry that the discussion of the problem of individuation is too complex to address in passing. Again, I am simply assuming that a proponent of the property view has a way to explain how a property is individuated such that it has only one object as a possible instance, e.g., through the introduction of the idea of *haecceities*. Alternatively, the property view could be reformulated in such a way that it is not one single or 'compound' property but a class or bundle of properties which comprises the properties whose instantiation is supposed to explain the existence of a particular object or individual.[5] But since nothing of my criticism hinges on the question of individuation or on the question if that whose having the property of being instantiated is a single property or a class or bundle of properties, I leave these issues aside. I present the property view as claiming that the existence of an object consists in a (single) property's having the property of being instantiated because I regard this as the least cumbersome way to express the core idea of this view.

3.2 Four Problems

3.2.1 A Vicious Circularity

The property view is unsatisfactory because it generates four problems. First, the view's explanation of what it is for an object to exist is viciously circular because a property's having the property of being instantiated presupposes that only existing objects are instances of the relevant property.[6] In other words, the existence of an object must be presupposed to make sense of the idea that a property has the property of being instantiated. A property's having the property of being instantiated requires that there is at least one instance of this property. But the idea that there is an instance of a property implies that this instance is an instance of the property because it exists. Something which does not exist cannot be an instance of a property. Thus, instantiation does not explain but presupposes existence.

Let us return to the example of Angela Merkel to illustrate the vicious circularity of the view. According to the property view, the existence of Angela Merkel

5 Cf. van Inwagen, *Metaphysics*, 297.
6 I am indebted to McGinn for the basic idea of this objection, see McGinn, *Logical Properties*, 21–23.

consists in a property's having the property of being instantiated. Let us assume that the relevant property is the property of being a woman named 'Angela Merkel' who was born on the 17th of July 1954 in Hamburg and who was the 8th chancellor of Germany after the Second World War.[7] If this is the case, the existence of Angela Merkel consists in this property's having the property of being instantiated.

We can ask the following question: What is required for a property to have the property of being instantiated? For example, what is required for the property of being a woman named 'Angela Merkel' who was born on the 17th of July 1954 in Hamburg and who was the 8th chancellor of Germany after the Second World War to have the property of being instantiated? This property can have the property of being instantiated only if there is an object which is an instance of this property and *by which* the property is instantiated. To be instantiated, this property and any other property must be instantiated *by* an object. But the property can be instantiated by an object only if this object exists.[8] In the case at hand, the property of being a woman named 'Angela Merkel' who was born on the 17th of July 1954 in Hamburg and who was the 8th chancellor of Germany after the Second World War is instantiated by Angela Merkel only if Angela Merkel exists.

Consequently, the explanatory order is inverse: It is not a property's having the property of being instantiated that explains the existence of an object. Rather, it is the existence of an object which instantiates a property which explains why a property has the property of being instantiated. Angela Merkel does not exist in virtue of the property of being a woman named 'Angela Merkel' who was born on the 17th of July 1954 in Hamburg and who was the 8th chancellor of Germany after the Second World War having the property of being instantiated. Rather, this property has the property of being instantiated in virtue of the existence of Angela Merkel.

To sum up: The answer of the property view to the question of what it is for an object to exist is viciously circular because this answer tries to identify and reduce the existence of an object to a property's having the property of being instantiated or having an instance. Then again, what is it for a property to have in-

7 Cf. ibid., 17–18; van Inwagen, *Metaphysics*, 297.

8 What about the possibility of combining the property view with a bundle theory? In this case, a property A's having the property of having an instance consists in a property B that exists to instantiate property A. I think not much is gained through this move because the objection could simply be reformulated in the appropriate way: The account of instantiation which is used to explain the existence of an object, i.e., a bundle of properties, is existentially loaded and has the idea that a property is instantiated only by properties that exist built into it. I am grateful to Jeremy Skrzypek for making me think about this possibility.

stances? A property's having the property of having an instance consists in an object that exists to instantiate the property.[9] Thus, the property view presupposes what is supposed to be explained by it, namely, the existence of an object. The account of instantiation which is used to explain the existence of an object is existentially loaded and has the idea that a property is instantiated only by objects that exist built into it.[10]

This can be shown with the example of the mistakenly posited inter-Mercurial planet Vulcan.[11] The decisive question is: Does the property of being a planet named 'Vulcan' which is in an orbit between Mercury and the Sun have the property of being instantiated? If the answer is yes, we have a good counterexample against the property view because there exists no such object. An object which is named 'Vulcan', and which is a planet in an orbit between Mercury and the Sun does not exist. If someone nevertheless wants to claim that Vulcan is an object which instantiates the property of being a planet named 'Vulcan' which is in an orbit between Mercury and the Sun, they must admit that a property's having the property of being instantiated has nothing to do with the existence of that object. Consequently, the view fails to explain what it is for an object to exist. The existence of an object cannot be identified with and reduced to a property's having the property of being instantiated because—as the example of Vulcan demonstrates—a property can be instantiated by an object like Vulcan which does not exist.

Consequently, the property view must negate the above-mentioned question: The property of being a planet named 'Vulcan' which is in an orbit between Mercury and the Sun does not have the property of being instantiated. Why not, then? What explains that this property does not have the property of being instantiated but that the property of being a planet named 'Mars' whose axial tilt is 25.19° relative to its orbital plane and whose orbital period is 687 Earth days does have the property of being instantiated? In other words, the property view must explain why the latter property has the property of being instantiated but not the former property.

Of course, proponents of the view could argue that only the latter property has the property of being instantiated because there is an object that exists by which the property is instantiated, namely, the planet Mars. Planet Mars is an object that exists, and therefore instantiates the property of being a planet named 'Mars' whose axial tilt is 25.19° relative to its orbital plane and whose or-

9 Cf. McGinn, *Logical Properties*, 23.
10 Cf. ibid., 22.
11 This example is taken from ibid.

bital period is 687 Earth days. The former property does not have the property of being instantiated because the planet Vulcan which is supposed to instantiate this property does not exist. Vulcan is not an object that exists, and therefore does not instantiate the property of being a planet named 'Vulcan' which is in an orbit between Mercury and the Sun.

If this is the explanation of a property's having the property of being instantiated, proponents of property view cannot argue that the explanation of what it is for an object to exist is that a property has the property of being instantiated. Such an explanation is viciously circular because the idea that only existing objects instantiate properties is used to explain what it is for a property to have the property of being instantiated. The existence of objects is not explained but rather presupposed by the view that the existence of an object consists in a property's having the property of being instantiated.

3.2.2 A Vicious Infinite Regress

The second problem with the property view is that it triggers a vicious infinite regress.[12] The infinite regress is triggered as soon as this view tries to explain what it is for the property to exist whose having the property of being instantiated is supposed to explain what the existence of an object consists in. Since the existence of an object is identified with and reduced to a property's having the property of being instantiated, this property must exist. It would be absurd to claim that the existence of an object X consists in a non-existing property's having the property of being instantiated. A property which does not exist cannot have the property of being instantiated and explain what the existence of an object consists in. In other words, something cannot exist in virtue of something else which does not exist. Thus, the property view must presuppose that the property, whose having the property of being instantiated is supposed to explain the existence of an object, exists itself.[13] For example, the property view's explanation of what it is for Angela Merkel to exist works only if the property of being a woman named 'Angela Merkel' who was born on the 17th of July 1954 in Ham-

12 This is a pretty standard objection. See, for example, ibid., 24–26. It could be argued that this objection is similar in nature to what is discussed under the name of 'Bradley's regress', see Katarina Perovic, "Bradley's Regress," ed. Edward N. Zalta, Winter Edition ed., *The Stanford Encyclopedia of Philosophy* (2017), https://plato.stanford.edu/archives/win2017/entries/bradley-regress/. For reasons of space, I will not investigate this question further. I am thankful to James Dominic Rooney for making me aware of this similarity.

13 Cf. McGinn, *Logical Properties*, 24.

burg and who was the 8th chancellor of Germany after the Second World War exists. For the sake of simplicity, let us call this property 'A'.

Now the following problem arises. How does the property view explain what it is to exist for A? A's existence must be explained because the existence of A plays a decisive role in the explanation of what it is for Angela Merkel to exist. So, to explain what it is for Angela Merkel to exist the view must explain what it is for a property like A to exist whose having the property of being instantiated is supposed to explain the existence of Angela Merkel. Let us consider our options.

A first option is to argue that the existence of A consists in A's having a property *sui generis*, namely, the property of existing. This option does not look very promising. First, *mutatis mutandis*, the property view's explanation of what it is for a property like A to exist would face the same problems as the non-reductive-object view. The same problems repeat themselves, only this time not at the level of objects but at the level of properties.

Second, if the property view claims that the existence of an object like Angela Merkel consists in a property A's having the property of being instantiated and that the existence of A consists in A's having a property *sui generis*, namely, the property of existing, this view becomes incoherent. On the one hand, the property view claims that the existence of an object can be identified with and reduced to *something else's* having a property, namely, a property's having the property of being instantiated. On the other hand, the property view claims that the existence of the property, whose having the property of being instantiated is supposed to explain the existence of the object, consists in that property's having a property *sui generis*, namely, the property of existing.

This second claim conflicts with the first because in the case of the property its existence is not identified with and reduced to *something else's* having a property. Thus, the idea that existence is something primitive which belongs to that which exists, is rejected at the level of objects but is reintroduced by the property view at the level of properties in order to for the property view to explain what it is for a property to exist whose having the property of being instantiated is supposed to explain the existence of an object. In short, the property view's non-reductive explanation of what it is for a property to exist conflicts with this view's reductive explanation of what it is for an object to exist. At the level of properties, the property view fails to provide a reductive explanation of what it is for a property to exist.

A defender of the property view could try to negate that this approach is incoherent. The property view could simply accept the idea that what it is for an object to exist is different from what it is for a property to exist. In other words, this view does not become incoherent by claiming that the existence of an object consists in a property's having the property of being instantiated and that the existence of that property consists in that property's having a property *sui generis*, namely, the property of existing, because objects and properties exist in different ways.

The problem is that this response makes the property view incoherent for another reason. The property view on what it is for an object to exist is usually considered to be committed to claims such as the following: that existence is what is expressed by the existential quantifier and only by it,[14] that existence is univocal,[15] and that the single sense of being or existence is adequately captured by the existential quantifier of formal logic.[16] But all of these claims conflict with the claim that objects and properties exist in different ways. If it is true that properties exist in a different sense or way than objects, it is false, for example, that the single sense of existence is adequately captured by the existential quantifier of formal logic.

A second option for the property view is to argue that the existence of the property A consists in A's having the property of being numerically identical with itself. This option is unattractive for the same reasons we have just seen: First, the problems which the reductive-object view has in explaining what it is for an object to exist, resurface—*mutatis mutandis*—at the level of properties. Second, if the property view claims that the existence of an object like Angela Merkel consists in a property A's having the property of being instantiated and that the existence of A consists in A's having the property of being identical with itself it becomes incoherent for the same reasons the first option becomes incoherent.[17]

The only way for the property view to remain coherent is to explain what it is for property A to exist in the same way as it explains what it is for an object to exist. Thus, to explain what it is for property A to exist, the property view must introduce another property B which has the property of being instantiated. And this property B cannot be any property, it must be a property of which A and only A is an instance. Otherwise, the existence of A could not be identified with and reduced to B's having the property of being instantiated.

Now we can see that this way of explaining what it is for an object like Angela Merkel to exist triggers a vicious infinite regress: The explanation of what it is for an object like Angela Merkel to exist requires the existence of property A which has the property of being instantiated by Angela Merkel and only by her. Yet, the explanation of what it is for A to exist requires the introduction

---

14 Cf. ibid., 20.

15 Cf. van Inwagen, "Being, Existence, and Ontological Commitment," 482–492.

16 Cf. ibid., 492–500.

17 For an additional problem with the view that the existence of a property consists in that property's having the property of being identical with itself, see McGinn, *Logical Properties*, 24, n.26.

of a further property B which has the property of being instantiated by A and only by A. The explanation of what it is for B to exist requires the introduction of a further property C which has the property of being instantiated by B and only by B and so on *ad infinitum.*

3.2.3 Ontological Inflationism

The third problem with the property view is that it works only with a heavily inflated ontology. If the existence of an object consists in a property's having the property of being instantiated, the view needs to introduce an infinite or at least very large number of properties which themselves exist in order to be able to explain the existence of the objects that exist.

As we have seen in the previous section, the property view needs to introduce an infinite number of existing properties into its ontology to explain what it is for Angela Merkel to exist. The same happens for every object that exists. Let us assume for the sake of argument that the property view is somehow able to be reformulated in such a way that its explanation of what it is for an object to exist does not trigger a vicious infinite regress. Thus, let us assume that the attempt to identify and reduce the existence of an object with and to a property's having the property of being instantiated is successful. In this case, all that is required to explain what it is for one object to exist is to introduce *one* property into one's ontology. For example, in the case of Angela Merkel, the property view needs to only argue that a property like being a woman named 'Angela Merkel' who was born on the 17th of July 1954 in Hamburg and who was the 8th chancellor of Germany after the Second World War exists and that this property has the property of being instantiated.

In this way, the problem of ontological inflationism is mitigated because the explanation of what it is for any object to exist does not require introducing an infinite number of existing properties into one's ontology. The problem is mitigated but not solved, though, because the property view must admit that it still needs a very large number of properties which exist themselves to be able to explain the existence of each object that exist. For each object that exists, there must be a different property which itself exists.

There must be a different property for each existing object because each property must individuate its object. Each property must be such that it is a property of which only one object is an instance, namely, the object whose existence is supposed to be explained. Otherwise, the existence of an object could not be identified with and reduced to the instantiation of this property. The existence of an object cannot be identified with and reduced to a property's having the property of being instantiated if the property can be instantiated by objects other than the object whose existence is supposed to be explained.

All these different properties must have existence themselves because it would be absurd to claim that the existence of an object is identified with and reduced to a non-existing property's having the property of being instantiated. Thus, even on the assumption that the property view's explanation of what it is for an object to exist does not trigger a vicious infinite regress, it works only with a heavily inflated ontology. The property view must introduce for each object that exists an existing property into its ontology which is such that it has only the relevant object as an instance.[18]

3.2.4 The Problem of the Co-Existence of Objects and Properties

A fourth and final problem with the property view is that it is confronted with the problem of the co-existence of objects and properties.[19] The problem arises because the property view must maintain that the properties—whose having the property of being instantiated is supposed to explain the existence of the objects that exist—exist themselves. Again, it would be absurd to claim that the existence of an object can be identified with and reduced to a non-existing property's having the property of being instantiated. The existence of something cannot be identified with and reduced to something which does not exist. Consequently, the property view is committed to an ontology which contains objects and properties, namely, those properties which individuate the existing objects and which have the property of being instantiated.

At this point the problem of their co-existence arises because this view cannot maintain that the properties—whose having the property of being instantiated is supposed to explain the existence of the objects that exist—exist *alongside* the ob-

18 What if the property view is combined with a bundle theory? In the case that no infinite regress is triggered, it seems that such a view is not vulnerable to the objection. Objects are just bundles of properties such that no new existing entities need to be introduced to account for the existence of objects. Maybe this works. But it seems to me that it is at best a trade-off. On the one hand, a bundle theory makes it possible to avoid the problem of ontological inflationism. On the other hand, a bundle theory is confronted with well-known problems of its own, e.g., to explain the identity and persistence of an object through time etc. Thus, although such a way to solve the problem of ontological inflationism seems viable, I believe it is not a very attractive option. I am thankful to Jeremy Skrypzek for making me aware of this possibility.

19 The problem of the co-existence of objects and properties is a problem which has certain similarities to what Huw Price and others have called a 'placement problem', see Huw Price, *Naturalism without Mirrors* (Oxford: Oxford University Press, 2011).

jects that exist. If this were the case, these properties would be just another type or kind of *objects.* Such objects would be a very strange kind of objects. These would be objects which—unlike any other existing object with which we are familiar with in our experience—can be instantiated. Furthermore, if these properties exist as a strange kind of objects alongside the objects whose existence they are supposed to explain, we get the wrong result if we count the *objects* which exist in mind-independent reality. For example, if the property of being a woman named 'Angela Merkel' who was born on the 17th of July 1954 in Hamburg and who was the 8th chancellor of Germany after the Second World War exists as a strange kind of object alongside Angela Merkel, there is not one object which exists in mind-independent reality but two *objects:* Angela Merkel and the property which accounts for her existence. This is absurd because it is not possible that there exists more than one object which is numerically identical with Angela Merkel.

Nor can the property view solve the problem of the co-existence of objects and properties by arguing that the properties *exist in* or *inhere in* the objects whose existence they are supposed to explain. This move conflicts with the reductive explanatory strategy of this view because properties which *exist in* or *inhere in* objects are properties of that object. Consequently, the existence of an object does not consist in *something else's* having a property but in that object's having a property. Thus, the property view collapses onto a version of the view that the existence of an object consists in that object's having a property. Furthermore, the strategy to solve the problem of co-existence by introducing different senses or modes of existing for objects and properties is not available for the property view because it makes this view incoherent for the reasons we have seen above (see 3.2.2).

As far as I can see, there remains only one way for the property view to solve the problem of the co-existence of objects and the properties whose having the property of being instantiated is supposed to explain the existence of the objects: The property view must divide mind-independent reality into different *ontological realms.* In other words, the property view must make use of the idea that there is a plurality of ontological realms. There is a realm in which objects exist and there is a realm in which the properties exist which account for the existence of objects. Thus, the property view must divide mind-independent reality into at least two ontological realms to be able to solve the problem of the co-existence of objects and the properties whose having the property to be instantiated is supposed to explain the existence of objects. [20]

20 An infinite number of ontological realms is required to solve the problem of co-existence if the problem of an infinite regress is not solved because the problem of co-existence repeats itself at the level of properties. For example, a property A whose having the property of being instantiated is supposed to explain the existence of an object requires the existence of a property B

It is important to note that I do not claim that this possible solution to the problem of co-existence makes the property view absurd. Like in the case of the problem of ontological inflationism, my aim is more modest. The point of these two objections is that the property view has considerable metaphysical costs. To explain what it is for an object to exist, the property view requires not only an ontology which contains an infinite or at least very large number of certain properties, but must divide mind-independent reality into an infinite number or at least two different ontological realms. So, ironically, the attempt to identify and *reduce* the existence of an object with and to a property's having the property of being instantiated works only through the *addition* of different kinds of entities and ontological realms which cannot be reduced to each other.

Especially the division of mind-independent reality into two or more different ontological realms makes the property view vulnerable to further objections. For example, it is notoriously difficult to explain how an external relation between an entity of one ontological realm, namely an object, and an entity of another ontological realm, namely a property, can account for a characteristic which the former entity has.[21]

Furthermore, if the existence of an object consists in its standing in an external relation to a property which exists in another ontological realm, it becomes difficult to understand how an object like Angela Merkel subsists or exists in itself in its own ontological realm. Not only is the subsistence of existing objects easily undermined. It also seems an inevitable consequence of this view that the ontological realm of the objects we are familiar with in our experience becomes a kind of second-class ontological realm because the existence of all these objects *depends* on certain properties which are members of a different ontological realm. There is no room here to elaborate on these issues. I only hint at some of them to indicate that the division of mind-independent reality into different ontological realms to solve the problem of co-existence produces a rat tail of other problems which are likely to make the property view even less attractive.

---

whose having the property of being instantiated is supposed to explain the existence of A. But B can neither exist alongside A nor exist in another sense or mode than A. Consequently, to solve the problem of the co-existence of A and B, another ontological realm must be introduced which contains properties like B but not properties like A and so on *ad infinitum*.

21 Cf. Koslicki, *Form, Matter, Substance*, 15–17; Edward J. Lowe, "A Neo-Aristotelian Substance Ontology: Neither Relational Nor Constituent," in *Contemporary Aristotelian Metaphysics*, ed. Tuomas Tahko (Cambridge: Cambridge University Press, 2012), 234–235.

4 The Unrestricted-Domain View

The last two views I am addressing in this first part of my book agree with the property view that the existence of an object X consists in something else Y's having a property. They part company with the property view when it comes to the question what kind of thing or entity Y is whose having a property is supposed to explain what it is for X to exist. The views I deal with in Chapters 4 and 5 argue that the existence of an object does not consist in a *property*'s having the property of being instantiated. Rather, the existence of an object X consists in the property of an unrestricted domain of objects—a UDO—or a restricted domain of objects—a REDO—of containing X and be [X]ish. In other words, the existence of an object X consists in a UDO's having the property of containing X and being [X]ish or a REDO's having the property of containing X and being [X]ish.[1] For the sake of convenience, let us call the latter view the restricted-domain view and the former view the unrestricted-domain view.

4.1 The Core Idea of the View

In this chapter, I start with a criticism of the unrestricted-domain view. A UDO is unrestricted in the sense that it is the domain of objects which contains *everything* that exists. Whatever a UDO is identified with—e.g., mind-independent reality, being, or the world—it is characterized as a *totality* which contains everything that exists.

In the contemporary debate, it has been argued that philosophers such as Kant, Milton Munitz, Bruce Aune, Nicolai Hartmann, and Frederic Sommers subscribe to versions of such a view.[2] Again, it is important to keep in mind that I am not concerned with the exegetical question whether it is true that any of the mentioned philosophers really entertain such a view. Rather, I address this view because it represents a systematic option in the debate and provides a possible answer to the nature question which is discussed by contemporary philosophers like William F. Vallicella or, more recently, Markus Gabriel.[3]

---

1 For the reconstruction of these views and in particular this way of formulating their core idea, I am indebted to Vallicella, *A Paradigm Theory of Existence*, 127–157.
2 Cf. Gabriel, *Sinn und Existenz*, 98–123, 138–139; Vallicella, *A Paradigm Theory of Existence*, 127.
3 Cf. Gabriel, *Sinn und Existenz*, 141–157, 224–270; Vallicella, *A Paradigm Theory of Existence*, 127–157.

https://doi.org/10.1515/9783110979879-007

The core idea of the unrestricted-domain view can be described in the following way:[4] There are objects which exist. For example, a quite plausible candidate for an object that exists is Angela Merkel. For the sake of argument, let us assume that it is true that Angela Merkel exists, i.e., that it is a *fact* that Angela Merkel exists. The following problem emerges. If it is a fact that Angela Merkel exists, it seems that it is a fact *about her.* This is the case because facts involve objects and the properties they have.[5] For example, the fact that Angela Merkel is brown-haired involves Angela Merkel and her property of having brown hair. In other words, what explains the fact that Angela Merkel is brown-haired is that having brown hair is her property. The problem is that the unrestricted-domain view cannot explain in the same way the fact that Angela Merkel exists because it rejects the idea that existence is a property of objects which exist.

If existence is not a property which Merkel has, how can the unrestricted-domain view explain that it is a fact that Angela Merkel exists? The central claim of this view is that we are mistaken to think that the fact that Angela Merkel exists is a fact *about her.* Rather, it is a fact *about something else.* It is important to distinguish *existential facts* (e.g., the fact that Angela Merkel exists) from *conventional facts* (e.g., the fact that Angela Merkel has brown hair). Existential facts are not facts about objects but facts about something else, namely, a UDO which contains everything that exists. Following a common usage, let us call this unrestricted domain of objects 'the world'.[6]

By introducing 'the world' as the unrestricted domain of objects which contains everything that exists, the unrestricted-domain view can explain the fact that an object X exists in the following way. The fact that an object X exists is not a fact in virtue of X's having the property of existing. Rather, the existence of X is a fact in virtue of the world's having the property of containing X, i.e.,

---

4 Here I am following rather loosely the way Vallicella explains the basic insight of the view which he calls the 'mondial attribute theory of existence', cf. *A Paradigm Theory of Existence*, 128–132.

5 This is a pretty standard way to characterize what a fact is, cf. McDaniel, *This Is Metaphysics*, 259. According to this way of characterizing what a fact is, the terms 'fact' and 'state of affairs' can be used interchangeably. It should be noted that in contemporary philosophy the questions of what a fact or state of affairs is and whether states of affairs are identical with facts is a matter of controversy, see Mark Textor, "States of Affairs," ed. Edward N. Zalta, Summer 2020 ed., *The Stanford Encyclopedia of Philosophy* (2020), https://plato.stanford.edu/archives/sum2020/entries/states-of-affairs/. The limited scope of this inquiry does not permit me to engage in this controversy and to justify the view on facts which I use here to explain the core idea of the unrestricted-domain view.

6 Cf. Gabriel, *Why the World Does Not Exist*; Vallicella, *A Paradigm Theory of Existence*, 128; van Inwagen, *Metaphysics*, 4, 291–292.

the world's being [X]ish.[7] For example, the explanation of the fact that Angela Merkel exists is not that she has the property of existing but that the world has the property of containing her, i.e., the world's being [Angela Merkel]ish.[8] Thus, there is a way to explain existential facts such as the fact that Angela Merkel exists without making use of the idea that existence is a property of that which exists.

It is important to note the reductive character of this strategy to explain existential facts. The explanatory strategy is structurally similar to the strategy which is employed by the property view. The existence of an object is identified with and reduced to a property which something else has. According to the unrestricted-domain view, the existence of an object X is identified with and reduced to the world's having the property of containing this X and being [X]ish.

Before I criticize this view, let us briefly consider what could be a motivation to prefer the unrestricted-domain view over the property view. The view that the existence of an object X consists in the world's having the property of containing X and being [X]ish has two advantages over the property view. First, the unrestricted-domain view's explanation of what it is for an object to exist works without a heavily inflated ontology. This approach is able to explain what it is for existing objects to exist without having to introduce an infinite or at least very large number of properties into its ontology which themselves exist. The unrestricted-domain view's explanation of what it is for that which exists to exist only requires that the world—understood as an unrestricted domain of objects, i.e., a UDO, which contains everything that exists—exists. The world must exist because it is the world's having the property of containing an object with which the existence of that object is identified with and reduced to. So, this view's explanation of what it is for an object to exist does not generate the problem of ontological inflationism and seems preferable to the property view due to its much more parsimonious ontology.

Second, the unrestricted-domain view's explanation of what it is for an object to exist seems simpler and more elegant than the explanation which the property view has to offer. Above we have seen (see 3.1) that the property view must claim that the property, whose having the property of being instantiated is supposed to explain the existence of an object, individuates this object. In other words, the property, whose having the property of being instantiated is supposed to explain the existence of an object, must be such that the object whose existence is supposed to be explained is the only possible instance of

7 Cf. Vallicella, *A Paradigm Theory of Existence*, 128.
8 Cf. ibid., 128, 135.

this property. Otherwise, the existence of an object cannot be identified with and reduced to the instantiation of this property.

Now, this requirement is not easy to meet, and things get really messy when the property view tries to explain what it is for a property A, whose having the property of being instantiated is supposed to explain what it is for an object to exist. The property view must assume that there exists a property B which individuates A, i.e., that there is a property which is such that property A is its only possible instance and so on.

The explanation which the unrestricted-domain view has to offer is less complicated and does not require that a UDO, i.e., the world, is such that it individuates the object whose existence it is supposed to explain. For example, the explanation of the fact that horses exist is that the world has the property of containing at least one horse, i.e., the world's being [horse]ish, and the explanation for the fact that unicorns do not exist is that the world lacks the property of containing at least one unicorn, i.e., the world's being un[unicorn]ish.[9] In a similar way, the unrestricted-domain view can explain the fact that an individual object like Angela Merkel exists. What explains the fact that Angela Merkel exists is that the world has the property of containing her, i.e., the world's being [Angela Merkel]ish. And what explains the fact that a fictional object like Sherlock Holmes does not exist is that the world lacks the property of containing Sherlock Holmes, i.e., the world's being un[Sherlock Holmes]ish. So, the existence and non-existence of objects consists in what the world contains and what it doesn't, i.e., how the world is and how it is not. The existence of object X consists in the world's being [X]ish and the non-existence of object Y consists in the world's being un[Y]ish.

4.2 Two Problems

4.2.1 A Vicious Circularity

The first problem with the view that the existence of object X consists in a UDO's having the property of containing X and being [X]ish is that it is viciously circular. For the sake of convenience, let us continue to call the relevant UDO, i.e., the unrestricted domain of objects which contains everything that exists, 'the world'. The unrestricted-domain view's explanation of what it is for an object to exist is viciously circular because the world's having the property of containing X and

---

9 Cf. ibid., 128.

being [X]ish presupposes that X exists. The existence of X must be presupposed to make sense of the idea that the world has the property of containing X and being [X]ish. Something which does not exist cannot be contained in the world. Thus, containment does not explain but rather presupposes the existence of that which is contained.

The vicious circularity becomes evident if we ask what is required for the world to have the property of containing X and being [X]ish. It seems that the world can have the property of containing X and being [X]ish only if X exists. But if this is the case, the world's having the property of containing X and being [X]ish does not explain what it is for X to exist because the existence of X—which is supposed to be explained by the world's having the property of containing X—is presupposed by the claim that the world has the property of containing X and is [X]ish. In other words, an account of what it is for the world to contain an object X and to be [X]ish presupposes an account of what it is for X to exist and can therefore not be used to explain what it is for an object X to exist. However, if the world's having the property of containing X and being [X]ish is *dependent* on the existence of X, the explanatory order is inverse. X does not exist in virtue of the world's having the property of containing X and being [X]ish. Rather, the world has the property of containing X and being [X]ish in virtue of the existence of X.

To sum up: The unrestricted-domain view's explanation that the existence of an object X consists in the world's having the property of containing X and being [X]ish is viciously circular because it presupposes that only existing objects are contained in the world. The account of containment which is used to explain what it is for an object X to exist is existentially loaded and has the idea that only objects that exist are contained in the world built into it.[10]

We can look the same point from a different angle. To evade this vicious circularity, the unrestricted-domain view must negate that the world can have the property of containing X and being [X]ish only if X exists. In other words, this approach must affirm that the world's having the property of containing X and being [X]ish is *not dependent* on the existence of X. Consequently, the unrestricted-domain view must affirm that it is possible that the world has the property of containing X and being [X]ish, even if X does not exist. Let us assume for the sake of argument that 'X' stands for the mistakenly posited inter-Mercurial planet Vulcan and ask the following question:[11] Does the world have the property of containing a planet named 'Vulcan', i.e., is the world [Vulcan]ish? If

10 Cf. McGinn, *Logical Properties*, 22.
11 This example is taken from ibid.

the answer is yes, we have a good counterexample against the claim that the existence of an object can be identified with and reduced to the world's property of containing this object because there exists no such planet. An object which is named 'Vulcan' and which is a planet in an orbit between Mercury and the Sun does not exist.

A proponent of the unrestricted-domain view can of course insist and somehow try to make sense of the claim that the world has the property of containing Vulcan or that the world is [Vulcan]ish. But regardless of how one makes sense of this claim, the world's having the property of containing Vulcan and being [Vulcan]ish has nothing to do with the existence of Vulcan. Consequently, the view fails to explain what it is for Vulcan to exist. The explanation cannot be that the existence of Vulcan consists in the world's having the property of containing Vulcan and being [Vulcan]ish because the world can have this property even if Vulcan does not exist.

It follows that the view's explanation of what it is for an object to exist works only if the question 'Does the world have the property of containing a planet named 'Vulcan', i. e., is the world [Vulcan]ish?' is negated and similar counterexamples are excluded. The unrestricted-domain view must negate that the world has the property of containing a planet named 'Vulcan' or other objects that do not exist. The world is not [Vulcan]ish.

The world does not have the property of containing Vulcan but does have the property of containing a planet named 'Mars' whose axial tilt is 25.19° relative to its orbital plane and whose orbital period is 687 Earth days. What is the explanation for that? Put simply: Why is the world not [Vulcan]ish but [Mars]ish? The unrestricted-domain view must explain why this is the case and on what grounds it attributes the property of containing Mars and being [Mars]ish to the world but not the property of containing Vulcan and being [Vulcan]ish. A proponent of the unrestricted-domain view could of course argue that the world has the property of containing Mars because Mars exists. The world has the property of containing Mars and is [Mars]ish in virtue of the existence of Mars. Accordingly, the world does not have the property of containing Vulcan and being [Vulcan]ish for the simple reason that Vulcan does not exist. There exists no such planet. Yet again, if this is the explanation for the world's having or lacking the property of containing an object X, i. e., the world's being [X]ish or not [X]ish, the view's explanation for the existence of X cannot be that X exists in virtue of the world's having the property of containing X and being [X]ish. Such an explanation would be viciously circular.

To sum up: The explanation of the unrestricted-domain view that the existence of an object X consists in the world's having the property of containing X and being [X]ish is viciously circular if the world's having the property of containing X and being [X]ish is *dependent* on the existence of X. However, if the world's having the property of containing X and being [X]ish is *not dependent* on the existence of X, the attempt to identify and reduce the existence of X with and to the world's having the property of containing X fails. It fails because there can be cases—like the case of Vulcan—where the world has the property of containing X and being [X]ish although X does not exist.

4.2.2 Incoherence

The second problem with the unrestricted-domain view is that it cannot even be formulated in a coherent way. On the one hand, this view must *affirm* that a UDO —e. g., the world—exists, i. e., it must affirm that there exists a domain of objects which contains everything that exists. The existence of an object X can be identified with and reduced to the world's property of containing X and being [X]ish only if the world itself exists. It would be absurd to claim that the existence of X consists in something Y's having the property of containing X and being [X]ish if Y does not exist. The existence of something cannot be identified with and reduced to a property of something else which does not exist. On the other hand, the unrestricted-domain view must *negate* that a UDO like the world exists. This view must negate that the world exists because, in its own terms, it is impossible for the world, i. e., for a UDO which contains everything that exists, to exist.

The argument runs like this:[12] As we have seen, the unrestricted-view's explanation for what it is for an object X to exist requires the existence of something like the world, i. e., a UDO which contains everything that exists. But what is it for the world to exist, i. e., what does the existence of the world consist in? According to the unrestricted-domain view, the existence of the world consists in a domain of objects' property of containing the world and being [world]ish. Let us call this domain of objects, which is supposed to explain what it is for the world to exist, 'super-world'. Super-world must exist because the existence of the world cannot be identified with and reduced to something which does not itself exist. It would be absurd to claim that the existence of the world consists in super-world's having the property of containing the world and being [world]ish if super-world does not exist. If there is a domain

12 A similar argument has been advanced recently by Markus Gabriel, whose view I will examine in the next chapter, see Gabriel, *Fields of Sense*, 187–209; *Sinn und Existenz*, 224–270; *Why the World Does Not Exist*, 83–98.

of objects which is such that it has the property of containing the world and being [world]ish, it must exist to be able to account for the existence of the world.

The problem with super-world is that it is impossible that it exists. Super-world is characterized as the domain of objects which has the property of containing the world which in turn is characterized as the unrestricted domain of objects which contains everything that exists. However, super-world cannot exist *and* have the property of containing the world because as soon as super-world exists there exists something which is not contained in the world, with the consequence that the world is not the world because it does not have the property of containing everything that exists.

In other words, as soon as there exists a domain of objects which is supposed to contain the world, it does not contain the world because it contains a domain of objects which does not contain everything that exists. The domain of objects which is supposed to contain the world does not contain the world because it does not contain the unrestricted domain of objects which contains everything that exists but only a restricted domain of objects. The domain of objects which is contained in super-world is restricted because it does not contain super-world which has the property of containing it. So, instead of a domain of objects which has the property of containing the world, there exists only a domain of objects which has the property of containing the domain of objects which contains everything that exists except the domain of objects which has the property of containing it.

To sum up: The unrestricted-domain view cannot be formulated in a coherent way because its explanation of what it is for an object X to exist requires the existence of something like the world, for which it is impossible to exist because it is not possible that there exists a further domain of objects which has the property of containing the world.

Let us briefly consider what possibilities there are to avoid this fatal result. To rescue the unrestricted-domain view from incoherence, one could try to argue that this view does not have to negate the existence of the world. The existence of the world does not have to be negated because the explanation of what it is for an object to exist does not apply to the world. Since the world is not an object but a domain of objects, what it is for the world to exist must be explained in a different way.

Such a move to safeguard the coherence of the unrestricted-domain view does not look not very promising for two reasons. First, to state that the existence of the world requires a different kind of explanation than the existence of an object is as such of no help. Since the existence of the world is presupposed by the view's explanation of what it is for an object to exist, the latter explanation is unsatisfactory or incomplete unless an explanation is given for what it is for the world to exist. In other words, the unrestricted-domain view cannot simply dissociate the problem of explaining what it is for the world to exist from the problem of explaining what it is for an object to exist.

Second, what theoretical options are there to explain what it is for the world to exist? As we have just seen, it is not an option for the unrestricted-domain view to explain the existence of the world in the way it explains the existence of an object. The existence of the world cannot be identified with and reduced to a domain of objects' property of containing the world and being [world]ish. Nor is it an option for this view to argue that the existence of the world consists in the world's having the property of existing. In this case, the unrestricted-domain view becomes vulnerable to all the objections which I have raised against the non-reductive-object view (see 1.2).

Furthermore, even if the unrestricted-domain view were able to rebut all these objections and defend the claim that the existence of the world consists in the world's having the property of existing, it is difficult to see what the motivation would be to explain in a different way what it is for an object to exist. Thus, if this view succeeds in defending the claim that the existence of the world consists in the world's property of existing, it most likely collapses into the view that the existence of an object consists in that object's having the property of existing. Moreover, the unrestricted-domain view becomes incoherent for another reason. On the one hand, this view claims that the existence of that which exists *can* be identified and reduced to something else's having a property. On the other hand, this view claims that existence of that which exists *cannot* be identified and reduced to something else's having a property.

5 The Restricted-Domain View

5.1 The Core Idea of the View

The last view I deal with in this first part of my inquiry maintains that the existence of an object X consists in the property of a restricted domain of objects—a REDO—of containing X and being [X]ish. In other words, the existence of an object X consists in a REDO's having the property of containing X and being [X]ish. My main interlocutor in this chapter will be Markus Gabriel, who recently presented and defended such a view.[1] For the sake of convenience, let us call this view the 'restricted-domain view', or alternatively 'Gabriel's view' without losing sight of the fact that similar positions have been recently advanced, for example, by authors like Jason Turner or Kris McDaniel.[2]

The core idea of this view comes into focus if contrasted with the unrestricted-domain view. For the sake of illustration, let us assume there are only three objects that exist, namely, the objects A, B, and C. And for the sake of convenience, let us again call the relevant UDO 'the world'. According to the latter view, the existence of A, B, and C consists in the world's having the property of containing A, B, and C and being [A]ish, [B]ish, and [C]ish. The existence of A, B, and C can be identified with and reduced to three different properties which

---

1 See, for example, "Existenz, realistisch gedacht."; *Fields of Sense*; "The Meaning of 'Existence' and the Contingency of Sense."; "Neutral Realism."; *Sinn und Existenz*; *Why the World Does Not Exist.* To be precise, Gabriel prefers to use a slightly different terminology. He calls a 'domain of objects' a 'field of sense'. Thus, in his terminology, his view is that the existence of an object X consists in a restricted field of sense's having the property of containing X and being [X]ish. He introduces the neologism 'field of sense' to differentiate his account of domains of objects from a mereological or set-theoretical interpretation of domains of objects and a merely extensionalistic understanding of the relation between an object and a domain of objects. But since nothing in my criticism hinges on this issue, I stick with the more familiar term 'domain of objects'. Furthermore, Gabriel frequently employs the verb 'to contain' to characterize his view but in general prefers the verb 'to appear'. Thus, in his terminology, for an object to exist is for it to appear in a field of sense. Again, by using 'to appear' instead of 'to contain' he wants to make it clear that the relation between an object and a domain of objects is not an extensionalistic relation like the relation between a part and a mereological sum or a member of a set and a set. But the use of the verb 'to appear' has other disadvantages, e.g., it invites subjective or anti-realist misconceptions of this view. Since nothing in my criticism presupposes a mereological or set-theoretical understanding of domains of objects, I stick with the verb 'to contain' to characterize the core idea of this view.

2 Cf. McDaniel, *The Fragmentation of Being*; Turner, "Ontological Pluralism."

https://doi.org/10.1515/9783110979879-008

the world has as the unrestricted domain of objects which contains everything that exists.

According to the restricted-domain view, the existence of an object X does not consist in the world's having the property of containing X and being [X]ish, i.e., in a UDO's having the property of containing X and being [X]ish, but in a REDO's having the property of containing X and being [X]ish. A REDO is restricted in the sense that it has the property of containing X but does not have the property of containing everything that exists. Thus, in our scenario, a *plurality* of restricted domains of objects is required to explain what it is for A, B, and C to exist.

There are at least two restricted domains of objects required to explain what it is for a plurality of objects to exist. For the sake of simplicity, let us assume that A, B, and C are contained in three different domains of objects. In this case, the existence of A consists in a $REDO_1$'s having the property of containing A and being [A]ish, the existence of B consists in a $REDO_2$'s having the property of containing B and being [B]ish, and the existence of C consists in a $REDO_3$'s having the property of containing C and being [C]ish. To sum up: The core idea of the restricted-domain view is that the existence of a *plurality* of objects cannot be identified with and reduced to a plurality of properties which *one* UDO has. Rather, the existence of a plurality of objects must be explained by a *plurality* of properties which a *plurality* of REDOs have.

Before I turn to a criticism of the restricted-domain view, let us briefly consider why one might want to embrace it. This question needs to be addressed because in comparison with the unrestricted-domain view, the restricted-domain view is clearly less elegant and has higher metaphysical costs due to its introduction of a plurality of domains of objects. For example, in our scenario, *three* domains of objects are required instead of *one* to explain what it is for A, B, and C to exist. Thus, there must be something which philosophers hope to get in return by embracing such a view.

As far as I can see, contemporary philosophers who adopt a version of the restricted-domain view are motivated by two different considerations. The main motivation for philosophers such as Jason Turner or Kris McDaniel to embrace such a restricted-domain view—also called 'ontological pluralism' in contemporary analytic philosophy—is the hope of rehabilitating the idea of modes of being or existence without having to give up a quantificational account of existence.[3] In other words, they hope that they can make sense of the idea that ob-

3 For some critical remarks on the prospects of McDaniel's version of the view, see Patrick Zoll,

jects exist in different ways with the help of a plurality of restricted quantifiers.[4] If there are different ways or modes of being or existing, one could argue, for example, that what it is for a concrete particular like a coffee cup to exist is not the same as what it is for an abstract universal like a number to exist.[5] Ontological pluralists like McDaniel argue that the attempt to rehabilitate the idea of modes of being or existence is worth the effort because with its help decisive contributions to ongoing debates in analytic metaphysics can be made.

What motivates Markus Gabriel to adopt a version of the restricted-domain view is not the aim to rehabilitate the idea of modes of being or existence. On the contrary, Gabriel rejects the claim that there are modes of being or existence.[6] Rather, what motivates Gabriel is the unsatisfactory answer which a quantificational account of existence and possible alternatives give to the question of what it is for an object to exist.[7] In particular, his aim is to disentangle the view that the existence of an object X consists in the property of a domain of objects of containing X and being [X]ish from the idea that the existence of an object X consists in a UDO's having the property of being [X]ish. With this strategy he hopes to achieve two things: First, the view that the existence of an object X consists in a REDO's having the property of containing X and being [X]ish can be formulated in a coherent way. Since a $REDO_1$, whose having the property of containing an object X and being [X]ish is supposed to explain the existence of X, is only a *restricted* domain of object, it is possible that there exists a further $REDO_2$ which has the property of containing $REDO_1$ and being [$REDO_1$]ish. Thus, the problem of incoherence is solved.

Second, the answer of the restricted-domain view to the nature question allows it to give quite a liberal and generous answer to the inventory question. In fact, Gabriel does not impose any restrictions on what could serve as a REDO other than that it is restricted, i.e., that it is a domain of objects which does not contain everything that exists. Whatever appears in a restricted domain of objects whatsoever, exists. Consequently, everything exists except the world and other objects which are such that it is impossible for them to appear in any REDO.[8] In this way, he believes to be able to successfully undermine any re-

"Review of Kris McDaniel, *The Fragmentation of Being*," *Zeitschrift für Theologie und Philosophie* 143, no. 1 (2021).

4 Cf. Cameron, "Critical Study of Kris McDaniel's *The Fragmentation of Being*," 785–788.

5 Cf. Turner, "Ontological Pluralism," 5.

6 Cf. Gabriel, *Fields of Sense*, 172–173.

7 Cf. ibid., 8, 43–186; Gabriel, *Sinn und Existenz*, 63–223.

8 Cf. *Fields of Sense*, 178.

ductive answer to the inventory question, especially naturalistic answers which deny the existence of the mind.[9]

5.2 Six Problems

5.2.1 A Vicious Infinite Regress

The first problem with the restricted-domain view is that it triggers a vicious infinite regress as soon as it tries to explain what it is for the REDO to exist whose having the property of containing an object X and being [X]ish is supposed to explain what it is for X to exist. To explain what it is for a $REDO_1$ to exist, the view must introduce another $REDO_2$ which has the property of containing $REDO_1$ and being [$REDO_1$]ish. At this point everything repeats itself because to explain what it is for $REDO_2$ to exist, the view must introduce a further $REDO_3$ which has the property of containing $REDO_2$ and being [$REDO_2$]ish and so on *ad infinitum.*

Let us take an example to illustrate why Gabriel's view triggers a vicious infinite regress. According to Gabriel, a spatio-temporal region like a room or a farm could serve as a restricted domain of objects.[10] Objects for which it is possible to appear in a spatio-temporal region are material objects like, for example, pigs or computers.[11] Now, let us assume that there exists a particular pig that is called 'Piggy' and that Piggy lives on a particular farm. Thus, the question is: 'What is it for Piggy to exist?'.

The answer which Gabriel's view has to offer is that the existence of Piggy consists in the farm's having the property of containing Piggy and being [Piggy]ish.[12] In other words, the existence of Piggy can be identified with and reduced to the farm's having the property of containing Piggy and being [Piggy]ish. The problem with this explanation is that it is not a satisfactory explanation of what it is for Piggy to exist. The explanation is not satisfactory because it presupposes but does not explain the existence of the farm whose having the property

---

9 Cf. ibid., 1–42; Gabriel, *Neo-Existentialism*, 8–45; "Neutral Realism."; *I Am Not a Brain: Philosophy of Mind for the Twenty-First Century* (Cambridge: Polity Press, 2017); "Für einen nichtnaturalistischen Realismus," in *Seien wir realistisch. Neue Realismen und Dokumentarismen in Philosophie und Kunst*, ed. Magdalena Marszalek and Dieter Mersch (Zürich; Berlin: Diaphanes, 2016); "Einleitung," 8–11, 16.

10 Cf. *Fields of Sense*, 60, 168.

11 Cf. ibid.

12 Cf. ibid., 60.

of containing Piggy and being [Piggy]ish is supposed to explain what it is for Piggy to exist. The farm must exist because it would be absurd to claim that the existence of Piggy consists in a non-existing farm's having the property of containing Piggy and being [Piggy]ish. Something which does not exist cannot have the property of containing existing Piggy. In other words, the existence of Piggy cannot be identified with and reduced to something Y's having a property if Y does not itself exist.

If the existence of the farm is presupposed by the restricted-domain view's explanation of what it is for Piggy to exist, the explanation of what it is for Piggy to exist is incomplete unless an explanation is given of what it is for the farm to exist. The farm plays an important role in the explanation of what it is for Piggy to exist, and the farm's existence cannot be taken for granted. It would be incoherent to claim that the existence of the farm can be taken for granted but that the existence of Piggy requires an explanation.

Since the farm which has the property of containing Piggy is itself only a restricted domain of objects there is no problem to account for what it is for the farm to exist. The existence of the farm consists in a further REDO's having the property of containing the farm. Expressed in Gabriel's terminology: The existence of a field of sense which has the property of containing an object consists in a further field of sense's having the property of containing the former field of sense.[13]

For the sake of argument, let us suppose the REDO which has the property of containing the farm which has the property of containing Piggy is the spatio-temporal region called 'Germany'. In this case, the existence of the farm consists in Germany's having the property of containing the farm and being [farm]ish. Thus, according to Gabriel's view, the existence of Piggy consists in the farm's having the property of containing Piggy and being [Piggy]ish whose existence consists in Germany's having the property of containing the farm and being [farm]ish. In other words, the existence of Piggy is identified with and reduced to Germany's having the property of containing the farm which has the property of containing Piggy.

At this point everything repeats itself because this explanation again is not a satisfactory explanation of what it is for Piggy to exist. The explanation is not satisfactory because it presupposes but does not explain what it is for the spatio-temporal region called 'Germany' to exist whose having the property of containing the farm is supposed to explain what it is to exist for the farm, whose having the property of containing Piggy and being [Piggy]ish is supposed to ex-

13 Cf. ibid., 225, 255.

plain what it is for Piggy to exist. The problem repeats itself *ad infinitum* because any further REDO which is introduced to explain what it is for Germany to exist requires the existence of a further REDO which has the property of containing Germany and so on.

This infinite regress which is triggered by the view's attempt to explain what it is for Piggy or any other object to exist is *vicious* because the infinite regress makes it impossible for the restricted-domain view to formulate an answer to the question what it is for Piggy or any other object to exist.[14] In other words, the infinite regress is vicious because the infinite character of the regress makes it impossible for the restricted-domain view to determine what that is which is supposed to explain the existence of an object. The central claim of the restricted-domain view is that the existence of an object X can be identified with and reduced to a property which something else Y has. On closer inspection, however, it results that this view is for a principled reason unable to determine what that property is and what the entity is which is supposed to have that property.

We get to the same conclusion via a different route. Gabriel formulates the core idea of his view sometimes in relational terms: The explanation for what the existence of an object X consists in is that a REDO stands in *relation* to X. Thus, the existence of X is identified with and reduced to a relation which obtains between a REDO and X, namely, the relation that a REDO contains X or that X appears in a REDO.[15]

If Gabriel's view wants to explain in this way what it is for an object X to exist, it is essential to determine both terms of the relation, i.e., to determine what the *relata* of the relation are. It is no problem to determine one *relatum*, namely, the *explanandum*. What is supposed to be explained is what it is *for X* to exist, for example, what it is *for Piggy* to exist. But the problem with the restricted-domain view is that it is unable to determine the other *relatum*, i.e., the *explanans* which is supposed to explain what it is for X to exist. Any REDO which has the property of containing X does not explain in a satisfactory manner what it is for X to exist because the existence of the REDO is presupposed, but it is not explained what it is for the REDO to exist.

As soon as the restricted-domain view tries to explain what it is for a REDO to exist a vicious infinite regress is triggered because the existence of a REDO_1 consists in a relation between a REDO_2 and the REDO_1, namely, the relation

14 For a similar problem in an epistemological context, cf. Paul A. Boghossian, *Fear of Knowledge: Against Relativism and Constructivism* (Oxford: Clarendon Press, 2006), 54–57.
15 Cf. Gabriel, *Fields of Sense*, 140, 267–268, 276; *Fiktionen*, 37–38, 75, 141, 181–182.

whereby a $REDO_2$ contains the $REDO_1$. The existence of the $REDO_2$ consists in a relation between a $REDO_3$ and the $REDO_2$, namely, the relation whereby a $REDO_3$ contains the $REDO_2$ and so on *ad infinitum.*

This infinite regress which is triggered by the view's attempt to explain what it is for an object X to exist is *vicious* because the infinite regress makes it impossible for the restricted-domain view to formulate a satisfactory answer to the question 'What is it for X to exist?': The existence of X consists in the relation which obtains between X and a $REDO_1$ (which contains X) and whose existence consists in the relation which obtains between the $REDO_1$ (which contains X) and a $REDO_2$ (which contains the $REDO_1$ which contains X) whose existence consists in the relation which obtains between the $REDO_2$ (which contains the $REDO_1$ which contains X) and a $REDO_3$ (which contains the $REDO_2$ which contains the $REDO_1$ which contains X) and so on *ad infinitum.*

Gabriel is aware of this problem and tries to solve it not by negating that an infinite regress is triggered by his way of explaining what it is for an object to exist but by negating that the regress is *vicious.*[16] Gabriel reconstructs the problem in a similar way as I have done above: Let d_1 be a restricted domain of objects, o_1 an object, R the relation of existence or actuality which obtains between an object and a restricted domain of objects, so that d_1Ro_1 is the relation of existence or actuality between d_1 and o_1.[17] A relation of existence or actuality between d_1 and o_1 obtains, if o_1 appears in d_1, i. e., if d_1 has the property to contain o_1.[18] However, if o_1 exists or is actual only as a *relatum* in the relation d_1Ro_1 and if this *relation* itself exists or is actual only if it appears as an *object* in some restricted domain of objects d_x, it seems that a vicious infinite regress of the form $d_\infty R\{...d_3R[d_2R(d_1Ro_1)_2]_3\}_\infty$ is triggered.[19]

Why? Because the relation d_1Ro_1 itself exists or is actual only if it appears as an object o_2 in the restricted domain of objects d_2. Now, as object o_2 the relation d_1Ro_1 exists or is actual only as a *relatum* in the relation of existence or actuality between d_1Ro_1 and a restricted domain of objects d_2 which is the relation $d_2R(d_1Ro_1)_2$. In short: The relation d_1Ro_1 exists or is actual only as a *relatum* of the relation $d_2R(d_1Ro_1)_2$.

16 Cf. *Fields of Sense*, 276–279.

17 Cf. ibid., 276–277. Gabriel uses the terms 'existence' and 'actuality' interchangeably. According to his view, the existence or actuality of an object X consists in the fact that X appears in a restricted domain of objects or in the fact that a restricted domain of objects has the property of containing X, cf. ibid., 264, 267.

18 Cf. ibid.

19 Cf. ibid., 140, 277.

Here, everything is repeated: The relation $d_2R(d_1Ro_1)_2$ itself exists or is actual only if it appears as an object o_3 in a restricted domain of objects d_3. As object o_3 the relation $d_2R(d_1Ro_1)_2$ exists or is actual only as a *relatum* in the relation of existence or actuality between $d_2R(d_1Ro_1)_2$ and d_3 which is $d_3R[d_2R(d_1Ro_1)_2]_3$ which itself exists or is actual only if it appears as an object o_4 in a restricted domain of objects d_4 and so on *ad infinitum*. Gabriel's example for such an infinite nesting of actualities is the following: "my hand is actual in the field of sense of my perception, which is actual in the field of sense of this phrase, which is actual in the field of sense of this book, which is actual in the field of sense of books you have read until page number XXX, and so on."[20]

Gabriel's strategy to deal with this problem is not to negate that such an infinite regress is triggered by his way of explaining what it is for an object to exist but to negate that the triggered regress is *vicious*.[21] How does he argue for this claim? His argument seems to be that an infinite regress and the resulting infinite nesting of actualities is only vicious if one accepts the additional premise that there is *one* ontologically privileged nesting of actualities which *connects* the different nestings of actualities to one great chain of being.[22] In other words, the infinite regress and the resulting infinite nesting of actualities are only vicious if the explanation of what it is for an object to exist presupposes that something like 'the world' exists, i.e., an unrestricted domain of objects which contains everything that exists or is actual.[23] Since he rejects the claim that the world exists and affirms the contrary, namely that there are not only infinitely *many* nestings of actualities but also indefinitely many nestings of actualities which are *not connected*, the regress is not vicious for his way to explain what it is for an object to exist.[24] In a nutshell: Gabriel argues that the problem of the viciousness of the infinite regress is solved by his rejection of the existence of the world and the acceptance of the claim that there are indefinitely many and not connected nestings of actualities.

This attempt to rebut the problem of a vicious infinite regress is flawed for two reasons. First, Gabriel's rebuttal misconstrues the objection by introducing the additional premise that there is *one* ontologically privileged nesting of actualities which *connects* the different nestings of actuality to one great chain of being. In other words, he smuggles into the objection the additional premise

20 Ibid., 277.
21 Cf. ibid.
22 Cf. ibid., 277–279.
23 Cf. ibid., 278.
24 Cf. ibid., 277–278.

that the world exists and then argues that an infinite regress is not vicious for his view on what it is for an object to exist because he rejects this premise.

This strategy is not legitimate because it shifts the topic and misconstrues the objection. The objection works without the assumption that there is a single ontologically privileged nesting of actualities which connects the different nestings of actualities to one great chain of being. In other words, since the problem of the viciousness of the infinite regress is not generated by the claim that the world exists, the problem is not solved by Gabriel's rejection of the existence of the world.

Second, Gabriel's rebuttal is flawed because it can be shown rather easily that the acceptance of the claim that there is an infinite plurality of different nestings of actualities which are not connected is of no help for Gabriel. The question of the number and the connectedness of the different nestings of actualities is simply irrelevant to the question whether the regress which is triggered by Gabriel's way of explaining what it is for an object to exist is vicious or not. As soon as Gabriel attempts to explain what it is for an object X to exist, an infinite regress is triggered which results in an infinite nesting of actualities. Such an infinite regress is vicious precisely because it results in an infinite nesting of actualities. The infinite character of the regress undermines the aim to explain what it is for an object to exist. Due to the triggered infinite regress, it is impossible for Gabriel's view to determine what that is which is supposed to explain what it is for an object to exist. The existence of an object X cannot be identified with and reduced to a property which something else Y has, because the restricted-domain view is for a principled reason unable to determine what that property is and what the entity is which is supposed to have that property.

What could Gabriel respond? I think the most promising strategy is to straightforwardly challenge the reason I have given for the viciousness of the infinite regress. I have argued that an infinite regress is vicious because it undermines the aim to formulate a satisfactory explanation of what it is for an object X to exist. Gabriel could try to challenge this argument in the following way: Let it be granted that his explanation of what it is for an object X to exist triggers an infinite regress which results in an infinite nesting of actualities because a $REDO_1$ which has the property of containing X must exist itself but can exist only if there is a further $REDO_2$ which has the property of containing the $REDO_1$ which has the property of containing X and so on *ad infinitum*. Thus, what is granted is that this way to explain what it is for X to exist inflates one's ontology with an indefinite number of restricted domains of objects. Gabriel could argue that this is a metaphysical cost or prize he is willing to pay but that this concession does not necessarily vitiate his explanation for what it is for X to exist. It does not do so because the view that the existence of an object X consists in a REDO's having the property of containing X and being [X]ish explains in a satisfactory way what it is for X to exist. Nothing more needs to be explained to satisfyingly explain what it is for X to exist.

For example, the existence of Piggy consists in a farm's having the property of containing Piggy and being [Piggy]ish. This explanation is a satisfactory explanation of what it is for Piggy to exist. It is of course true that this explanation presupposes the existence of the farm but does not explain what it is for the farm to exist. But to explain what it is for the farm to exist is different story. The restricted-domain view has no problem with explaining in a satisfactory manner what it is for the farm to exist. For example, Gabriel could argue that the existence of the farm consists in another REDO's having the property of containing the farm and being [farm]ish (e.g., the spatio-temporal region called 'Germany'). In this way, a satisfactory explanation of what it is to exist can be given for each member of an infinite nesting of actualities.

To sum up: Gabriel's explanation of what it is for an object X to exist implies that X is embedded in a $REDO_1$ which in turn is embedded as an object in a further $REDO_2$ which in turn is embedded as an object in a further $REDO_3$ and so on *ad infinitum*.[25] Gabriel could try to argue that this infinite regress is not vicious because what it is for X to exist is satisfactorily explained through a $REDO_1$'s having the property of containing X, what it is for the $REDO_1$ to exist is satisfactorily explained through a $REDO_2$'s having the property of containing the $REDO_1$ and so on *ad infinitum*. Thus, it seems that my argument for the vicious character of the infinite regress is successfully rebutted by Gabriel.

First, it is possible to determine a REDO whose having the property of containing an object X explains what it is for X to exist. For example, in the case of Piggy, it is the farm's having the property of containing Piggy and being [Piggy]ish which explains what it is for Piggy to exist. And in the same way there is no principled problem with determining for every REDO within an infinite nesting of actualities a further REDO which has the property of containing the former.

Second, since the *explanandum* and the *explanans* can be determined, it is also possible to determine the property or relation with which the existence of an object X is identified with and reduced to. The existence of X is identified with and reduced to the relation which obtains between a REDO and X, namely, the relation that a REDO contains X or that X appears in a REDO. For example, the existence of Piggy consists in the relation which obtains between Piggy and the farm, namely, that the farm contains Piggy or that Piggy appears in the farm. In the same way there is no principled problem with identifying and

25 Cf. ibid., 158.

reducing the existence of every REDO within an infinite nesting of actualities to a relation which obtains between a REDO and a further REDO.

The crucial problem with this rebuttal is that it slightly but decisively misconstrues the problem of an infinite nesting of actualities, i.e., the problem of an infinite embeddedness of restricted domains of objects in further restricted domains of objects. How Gabriel misconstrues the problem can be illustrated by using his example for an infinite nesting of actualities: "my hand is actual in the field of sense of my perception, which is actual in the field of sense of this phrase, which is actual in the field of sense of this book, which is actual in the field of sense of books you have read until page number XXX, and so on."[26]

At a first glance, this looks like an example of an infinite nesting of actualities, i.e., an infinite embeddedness of restricted domains of objects in further restricted domains of objects. However, it is not, because it does not take into account Gabriel's commitment to the claim that an object is a *relatum* which exists only in a relation to a restricted domain of objects which has the property of containing it.[27] Due to this commitment, the correct formulation of the example has to be: my hand is actual in the restricted domain of objects which is my perception, my perception *in which my hand is actual* is actual as an object in the restricted domain of objects which is this phrase, this phrase *in which my perception is actual in which my hand is actual* is actual as an object in the restricted domain of objects which is this book and so on *ad infinitum*. Such an infinite regress is vicious for the aim to explain what it is for an object to exist because the explanation never comes to an end. For a principled reason, we never get a satisfactory explanation of what it is for an object to exist.

We can also use Gabriel's own logical reconstruction for an infinite nesting of actualities, i.e., the infinite embeddedness of restricted domains of objects in further restricted domains of objects to explain why Gabriel's example must be corrected as proposed:

$$d_\infty R \{ \ldots d_3 R[d_2 R(d_1 R o_1)_2]_3 \}_\infty.$$ [28]

According to Gabriel, for his hand to exist is to appear in the restricted domain of objects which is his perception. The existence of his hand can be identified with and reduced to his perception's having the property of containing his hand, i.e.,

26 Ibid., 277.
27 Cf. ibid., 140, 267–268; Gabriel, *Fiktionen*, 37–38, 75, 141, 181–182.
28 Cf. *Fields of Sense*, 277.

it can be identified with and reduced to the relation that his perception contains his hand or the relation whereby his hand appears in his perception.[29]

Thus, for o_1 (Gabriel's hand) to exist is to appear as a *relatum* in the relation d_1Ro_1 (Gabriel's perception in which his hand appears). This explanation for what it is for o_1 (Gabriel's hand) to exist is not satisfactory because it presupposes the existence of the restricted domain of objects d_1 (Gabriel's perception). Gabriel's perception is the other *relatum* of the relation d_1Ro_1 (Gabriel's perception in which Gabriel's hand appears) which must exist because the existence of d_1 is constitutive of the existence of the relation d_1Ro_1. Without d_1 there is no relation d_1Ro_1 which is supposed to explain what it is for o_1 to exist.

Gabriel seems to believe that this explanation for what it is for o_1 to exist does not trigger a *vicious* infinite regress because the existence of o_1 (Gabriel's hand) is satisfactorily explained through the relation d_1Ro_1 (Gabriel's perception in which Gabriel's hand appears). It is true, Gabriel claims, that this way of explaining what it is for o_1 to exist triggers an infinite regress because it requires the existence of d_1 (Gabriel's perception) which requires the existence of a further restricted domain of objects which in turn requires the existence of a further domain of objects and so on *ad infinitum.*[30] However, such an infinite regress is not vicious for the explanation of what it is for o_1 (Gabriel's hand) to exist because what it is for d_1 (Gabriel's perception) as o_2 to exist can be satisfactorily explained by the appearing of o_2 in the restricted domain of objects d_2 which is this phrase, i.e., this phrase's having the property of containing Gabriel's perception.[31]

I think, at this point, one should be able to see what went wrong with Gabriel's example and his attempt to rebut the objection that the infinite regress is vicious which is triggered by his way of explaining what it is for an object like his hand to exist. What appears as o_2 in d_2 (the restricted domain of objects

---

29 I think Gabriel's example is problematic for other reasons as well. For example, it seems absurd to claim that the existence of Gabriel's hand consists in the property of Gabriel's perception to contain his hand. His hand does not cease to exist if his perception ceases to have the property of containing his hand. This problem is not restricted to this example but is a general problem with his view, which I will deal with below (see 5.2.4). In this context, I am only interested in showing why Gabriel's attempt to rebut the objection that the infinite regress is vicious fails. For this reason, I ignore such additional problems with his example.

30 Cf. Gabriel, *Fields of Sense*, 159–160.

31 Again, Gabriel's example is problematic because it seems rather absurd to claim that the existence of Gabriel's perception consists in a phrase's having the property of containing his perception. It is pretty unintelligible how anybody's perception can be contained by a phrase in such a way that the existence of that perception consists in the phrase's property of containing the perception. But as I said before, I ignore this problem here because I am only interested in showing why Gabriel's attempt to rebut the objection that the infinite regress is vicious fails.

which is this phrase) is not just d_1 (Gabriel's perception) but d_1Ro_1 (Gabriel's perception *in which his hand appears*). Thus, Gabriel misconstrues the *object* whose existence needs to be explained. What needs to be explained to have a satisfactory explanation of what it is for o_1 (Gabriel's hand) to exist is what it is for d_1Ro_1 (Gabriel's perception in which his hand appears) as o_2 to exist. The relation d_1Ro_1 as o_2 is an object with an inner structure which is constituted by the *relata* d_1 and o_1. For this reason, what it is for d_1Ro_1 (Gabriel's perception in which his hand appears) to exist is not sufficiently explained by the relation which obtains between d_1 (Gabriel's perception) and d_2 (the restricted domain of objects which is this phrase) because this explains at best what it is for d_1 (Gabriel's perception) to exist but not what it is for d_1Ro_1 (Gabriel's perception in which his hand appears) to exist.

Therefore, as soon as Gabriel attempts to explain what it is for the relation d_1 Ro_1 (Gabriel's perception in which his hand appears) to exist his explanation for what it is for o_1 (Gabriel's hand) to exist becomes vicious because the relation d_1 Ro_1 as object o_2 only exists itself as *relatum* in the relation which obtains between o_2 and d_2 (the restricted domain of objects which is this phrase) which is $d_2R(d_1$ $Ro_1)_2$ (the phrase in which Gabriel's perception appears in which his hand appears) and so on *ad infinitum*. To sum up: The regress which is triggered by Gabriel's way to explain what it is for an object X to exist is vicious and his attempt to show otherwise fails.

5.2.2 Ontological Inflationism

The second problem with the restricted-domain view is that it only works with a heavily inflated ontology. As we have seen in the previous section, this view triggers an infinite regress as soon as it tries to explain what it is for the REDO to exist whose having the property of containing X and being [X]ish is supposed to explain what it is for an object X to exist. The existence of X requires the existence of a $REDO_1$ which has the property of containing X. The existence of the $REDO_1$ requires the existence of a $REDO_2$ which has the property of containing the $REDO_1$ which has the property of containing X. The existence of the $REDO_2$ requires the existence of a $REDO_3$ which has the property of containing the $REDO_2$ which has the property of containing the $REDO_1$ which has the property of containing X and so on *ad infinitum*. Consequently, the restricted-domain view requires the existence of an infinite number of restricted domains of objects to be able to explain what it is for just *one* singular object X to exist.

It is important to note that the problem of ontological inflationism is distinct from the problem of the *viciousness* of the infinite regress. The discussion in the previous section was about the question if the triggered infinite regress is vicious for the aim of the restricted-domain view to explain what it is for an object X to exist. Now, this view faces the problem of ontological inflationism regardless of whether the infinite regress is vicious or not. Even if Gabriel could somehow show that the triggered infinite regress is not vicious for his explanation of what it is for an object X to exist, he is still confronted with the problem that an infinite regress inflates this view's ontology to the point of absurdity. To evade this consequence, Gabriel would have to deny that an infinite regress is triggered by his way to explain what it is for an object X to exist. But he does not. On the contrary, he openly admits that an infinite regress is triggered and that this infinite regress goes hand in hand with an infinite inflation of the ontology he is willing to embrace.[32]

Thus, my objection in this section is simply that Gabriel's answer to the nature question makes the restricted-domain view's answer to the inventory question (i.e., the question of what there is or exists) highly unattractive. In other words, the answer of Gabriel's view to the nature question has a significant metaphysical cost or price. The restricted-domain view on what it is for an object to exist requires the existence of an infinite number of restricted domains of objects. Without the existence of an infinite number of restricted domains of objects, this view is unable to explain what it is for just *one* object to exist. Thus, if there is a view on what it is for an object to exist which works with a more parsimonious ontology, it is clearly preferable to the restricted-domain view.

5.2.3 A Vicious Circularity

The third problem with the restricted-domain view is that its explanation for what it is for an object X to exist is viciously circular. It is viciously circular because any REDO's having the property of containing X presupposes that X exists. The existence of X must be presupposed to make sense of the idea that a REDO has the property of containing X. Something which does not exist cannot be contained by a REDO. Thus, containment does not explain but rather presupposes the existence of that which is contained.

The vicious circularity becomes evident if we ask what is required for a REDO to have the property of containing X and being [X]ish. It seems that a

32 Cf. Gabriel, *Fields of Sense*, 159–160, 196, 242; "The Meaning of 'Existence' and the Contingency of Sense," 127–128.

REDO can only have the property of containing X and being [X]ish if X exists. If this is the case, however, a REDO's having the property of containing X and being [X]ish does not explain what it is for X to exist because the existence of X—which is supposed to be explained through the REDO's having the property of containing X—is presupposed by the claim that a REDO has the property of containing X and is [X]ish. In other words, an account of what it is for a REDO to contain an object X and to be [X]ish presupposes an account of what it is for X to exist and can therefore not be used to explain what it is for an object X to exist.

If a REDO's having the property of containing X and being [X]ish is *dependent* on the existence of X, the explanatory order is inverse. X does not exist in virtue of a REDO's having the property of containing X and being [X]ish. Rather, a REDO has the property of containing X and being [X]ish in virtue of X's existence. To sum up: The restricted-domain view's explanation of what it is for an object X to exist is viciously circular because it presupposes that X exists. The account of containment, which is used to explain what it is for an object X to exist, is existentially loaded, and has the idea built into it that X exists.[33]

The vicious circularity can be illustrated by using one of Gabriel's examples. The scenario is as follows: Gabriel is sitting in a room in San Francisco with a mug full of coffee.[34] Now, according to Gabriel's view, for an object to exist is to appear in restricted domain of objects (a field of sense in his terminology), so that for the coffee to exist is to appear in the mug.[35] The existence of the coffee can be identified with and reduced to the mug's having the property of containing the coffee and being [coffee]ish.

The circularity of this explanation becomes evident as soon as we ask what is required for the mug to have the property of containing the coffee and being [coffee]ish. It seems clear that the mug can have the property of containing coffee and being [coffee]ish only if there is coffee in it. Yet there can be coffee in it only if the coffee which is in it exists. If the coffee which is in the mug does not exist, there is no coffee at all in the mug. However, if this is the case, the mug's having the property of containing the coffee and being [coffee]ish does not explain what it is for the coffee to exist because the existence of the coffee—which is supposed to be explained through the mug's having the property of containing the coffee—is presupposed by the claim that the mug has the property of containing the coffee. In other words, to make sense of the claim that the mug has the property of

33 Cf. McGinn, *Logical Properties*, 22.
34 Cf. Gabriel, *Fields of Sense*, 160.
35 Cf. ibid., 44, 101, 140, 158, 160, 276–277; Gabriel, *Sinn und Existenz*, 77–78, 224.

containing the coffee and being [coffee]ish, the existence of the coffee and consequently an account of what it is for the coffee to exist must be presupposed and can therefore not be used to explain what it is for the coffee to exist.

Yet, if this is the case, the mug's having the property of containing the coffee is *dependent* on the existence of the coffee with the consequence that the explanatory order is inverse. The coffee does not exist in virtue of the mug's having the property of containing the coffee and being [coffee]ish. Rather, the mug has the property of containing the coffee and being [coffee]ish in virtue of the existence of the coffee in it.

The only way to avoid the circularity would be to claim that the mug's having the property of containing the coffee is *not dependent* on the existence of the coffee in it. Thus, the mug can have the property of containing coffee without it being the case that it contains coffee which exists. There are two problems with this tactic. First, it is unintelligible how the mug can have the property of containing coffee without it being the case that the coffee which it contains exists. Second, if containment is not dependent on the existence of that which is contained, we are free to claim that Gabriel's mug also has the property of containing tea and being [tea]ish. In this case, the existence of the coffee cannot be identified with and reduced to the mug's having the property of containing the coffee and being [coffee]ish because the mug can have the property of containing something without it being the case that that which it contains exists.

To sum up: The explanation that the existence of an object X consists in a REDO's having the property of containing X and being [X]ish is viciously circular, if the REDO's having the property of containing X and being [X]ish is *dependent* on the existence of X. However, if a REDO's having the property of containing X and being [X]ish is *not dependent* on the existence of X, the attempt to identify and reduce the existence of X with and to a REDO's having the property of containing X fails. It fails because there can be cases where a REDO has the property of containing X and being [X]ish although X does not exist.

5.2.4 The Problem of Constitution

The fourth problem with the restricted-domain view is that a REDO's having the property of containing X and being [X]ish is not constitutive of the existence of X because X can exist without being contained by the REDO or without appearing in the REDO. To illustrate this problem, let us return to the example of the previous section, namely, the coffee in Gabriel's mug.[36] I think it is plausible that the mug's having the property of containing the coffee and being [coffee]ish is not constitutive of the existence of the coffee. For example, if Gabriel decides to go for a walk and pour the coffee which is in his mug into a thermos flask, his mug has lost the property of containing coffee and being [coffee]ish, but the coffee has not ceased to exist. Rather, it now exists in the thermos flask. Consequently, the existence of the coffee cannot be identified with and reduced to the mug's having the property of containing coffee and being [coffee]ish. Evidently, the existence of the coffee does not consist in the mug's having the property of containing the coffee and being [coffee]ish because the coffee can exist without it being the case that it is contained in the mug or appears in the mug.

Even worse, if there were a REDO, whose having the property of containing the coffee is constitutive of the existence of the coffee, it would result that the coffee does not exist as what we believe it to exist, namely, as an object which exists in its own right, i.e., as an object which exists in itself.[37] If there exists a restricted domain of objects, whose having the property of containing the coffee and being [coffee]ish is constitutive of the existence of the coffee, the coffee does not exist as an object which exists in itself, i.e., as an object that exists in its own right. Rather, in this case, the coffee exists either as an object which exists in the relevant REDO as a part of it or as an object which inheres in the relevant REDO as an accidental property of it. In other words, if there exists a REDO, whose having the property of containing the coffee and being [coffee]ish is constitutive of the existence of the coffee, the coffee exists not in itself and in its own right but either exists in the REDO as a part of it or inheres in the REDO as an accident of it.

The first option has the very problematic consequence that even objects which we consider to be paradigmatic examples of objects which exist in their own right (e.g., persons like Angela Merkel or you and me) lose that status and become objects which only exist as parts in larger wholes. Furthermore, if such objects exist as parts in wholes, the relation between an object and a restricted domain of objects, which is supposed to explain what the existence of that object consists in, becomes a mereological relation. Such a mereological

---

36 Cf. *Fields of Sense*, 160. Other examples of Gabriel's which could be used to make the same point are a computer in a room, water in a Fata Morgana, or a hand in a visual field, cf. ibid., 168–169, 277.

37 In case you doubt that coffee is an object which exists in its own right, feel free to substitute the coffee with Angela Merkel, yourself, or another object which you believe exists in its own right.

or extensionalist interpretation of the existence relation is rejected by Gabriel.[38] Thus, the first option is not available for Gabriel and he must adopt the second option.

The second option has even more drastic consequences than the first. According to the second option, objects like coffee, Angela Merkel, or you and me, do not exist in their own right but exist only in another as a subject, i.e., as a modification of a restricted domain of objects. For example, the coffee exists only as a modification of the mug, i.e., as an accidental form of the mug in virtue of which the mug is [coffee]ish. Piggy exists only as a modification of a farm, i.e., as an accidental form of a farm in virtue of which the farm is [Piggy]ish.

Thus, Gabriel's view does not only fail to account for the existence of objects which exist in themselves. It also actively undermines the existence of such objects due to its commitment to the claim that the existence of any object consists in a REDO's having the property of containing that object. Consequently, any object X exists only as a modification of a REDO. In other words, Gabriel's view has not only the problem that it cannot account for the existence of substances but also that it 'accidentalizes' substances, i.e., that it transforms substances into accidents.

What's worse, Gabriel's view cannot even be formulated in an intelligible way because the problem repeats itself at the level of any REDO whose having the property of containing an object X is supposed to explain what it is for an object to exist. According to Gabriel's view, the existence of any $REDO_1$ consists in a further $REDO_2$'s having the property of containing the $REDO_1$ and being [$REDO_1$]ish. Thus, any $REDO_1$ is not an object which exists in itself and in its own right. Rather, any $REDO_1$ is either an object which exists in a $REDO_2$ as a part of it or is an object which exists in a $REDO_2$ as an accidental form of it in virtue of which the $REDO_2$ is [$REDO_1$]ish. The first option must be ruled out for the reasons given above, so that Gabriel's view must claim that any $REDO_1$ exists only as a modification of a $REDO_2$ which exists only as a modification of a $REDO_3$ and so on *ad infinitum*. A view which states that there are no objects which exist in themselves, i.e., in their own right, but only infinite chains of modifications of modifications is unintelligible.

I think Gabriel commits the fallacy of inferring from the true *general* claim that any object X that exists is embedded in a context that a *particular* context

38 Cf. Gabriel, *Fields of Sense*, 12, 94–95, 160, 229; "Repliken auf Diehl/Rosefeldt, Hübner, Rödl und Stekeler-Weithofer," in *Jahrbuch-Kontroversen 2: Markus Gabriel – Neutraler Realismus*, ed. Thomas Buchheim (Freiburg; München: Verlag Karl Alber, 2016), 189–192.

in which X is embedded is constitutive of the existence of X.[39] Gabriel's fallacy is structurally similar to the fallacy of inferring from the true *general* claim that any material object X has accidents or accidental properties—which, for example, determine X with respect to its quality, quantity, or location— that it is constitutive for X to have any *particular* accidents. In other words, it is true that any material object necessarily has accidents or accidental properties, but it is false that it is necessary that it has one accident or accidental property rather than another.[40] In an analogous way, it is true that any object that exists is necessarily embedded in a context.[41] But it is fallacious to infer from this true claim that it is necessary that any object is embedded in a *particular* context.

5.2.5 Reductionism Does Not Work

The fifth problem with Gabriel's view is that its attempt to identify and reduce the existence of an object X with and to a REDO's having the property of containing X and being [X]ish does not work.[42] The reductionism does not work because any object that exists exists as an object which has accidental properties and Gabriel's view cannot account for this fact.[43]

To illustrate why Gabriel cannot explain what it is for an object with accidental properties to exist, let us suppose that Joe Biden exists and that he is sitting

---

39 Cf. *Fields of Sense*, 158; *Sinn und Existenz*, 224–225. To anticipate the objection that God might be an object that exists but which is not embedded in a context, it would be more precise to say that any object that *has existence* is embedded in a context. But the existence of God is a special case which is not my concern here, so I leave out this complication.

40 Cf. Stump, *Aquinas*, 50.

41 Again, it would be more precise to say that any object that necessarily *has existence* is embedded in a context to rule out the possible counterexample of God. But as I said above, I am not concerned here with the possible exceptional case of God but with existing objects we are familiar with in our experience, e.g., coffee, mugs, animals, or human beings. To evade unnecessary complications, I bracket off the issue of the possible existence of God.

42 The idea for this objection is taken up from Vallicella but I do not follow exactly his way of formulating the objection, see Vallicella, *A Paradigm Theory of Existence*, 132–135. The objection applies also to the unrestricted-domain view.

43 Again, it is important to note that if God exists as an entity whose essence is identical with its existence, he is an exception because he is an object that exists, but which does not have any accidental properties. An entity whose essence is identical with its existence is metaphysically simple, i.e., it is impossible that it has accidental properties because having accidental properties requires a metaphysical composition. For a more detailed argument, see, for example, *ST*, I, q. 3, a. 6. I bracket the issue of the possible existence of God to avoid making my argument more complicated than necessary.

behind his desk in the White House at this moment. In this case, Biden exists as a sitting human being, i.e., as a human being which possesses the accidental property of sitting. The following problem arises: How does Gabriel's view explain what it is for *sitting* Joe Biden to exist?

The explanation that the existence of Biden consists in the White House's having the property of containing Biden and being [Joe Biden]ish is not satisfactory because this explanation only explains what it is for *Joe Biden* to exist but not what it is for *sitting* Joe Biden to exist. Furthermore, Gabriel cannot argue that the existence of sitting Joe Biden consists in Joe Biden's having the accidental property of being sitting because Joe Biden's having the accidental property of being sitting does not explain what it is for Joe Biden to exist but presupposes the existence of Biden. Thus, to explain what it is for *sitting Joe Biden* to exist, Gabriel's view has to argue that the existence of sitting Joe Biden consists in the White House's having the property of containing sitting Joe Biden and being [sitting Joe Biden]ish.

The problem with this explanation is that the existence of sitting Biden cannot be identified with and reduced to a property which something else has, namely, the White House. The reduction does not work because Joe Biden's having the property of being sitting cannot be identified with and reduced to a property which the White House has. It would be absurd to claim that the White House has the property of being sitting. Rather, if the White House has the property of containing *sitting* Joe Biden, it is in virtue of Joe Biden's having the property of being sitting. Yet Joe Biden can have the property of being sitting only if he exists. To sum up: The existence of an object X that has accidental properties (e.g., sitting Joe Biden) cannot be identified with and reduced to a REDO's having the property of containing X and being [X]ish because having these accidental properties cannot be identified with and reduced to properties which the REDO has.

The same point can be made from a slightly different angle.[44] The restricted-domain view is a variant of the general view the core idea of which can be expressed as follows: The fact that an object like Joe Biden exists is not a fact *about him* but a fact about *something else*, namely, a domain of objects (see 4.1). In other words, it is central for Gabriel's view to distinguish between *existential facts* (e.g., the fact that Joe Biden exists) and *conventional facts* (e.g., the fact that Joe Biden is sitting). An existential fact like the existence of Joe Biden is not a fact in virtue of Joe Biden's having the property of existing. Rather,

44 Cf. Vallicella, *A Paradigm Theory of Existence*, 128–137.

the existence of Joe Biden is a fact in virtue of a REDO's (e.g., the White House) having the property of containing him and being [Joe Biden]ish.

This explanatory strategy is reductive in character because the existence of an object like Joe Biden is identified with and reduced to a property which something else has, namely, a restricted domain of objects like the White House. The example of sitting Joe Biden makes it evident that this reductive explanatory strategy does not work because it presupposes that the existential fact that Joe Biden exists can be separated from all conventional facts about Joe Biden. But it cannot. There are no purely existential facts.[45] An object X's 'having' an accidental property like sitting is something which cannot be identified with and reduced to a property something else Y has. The fact that Joe Biden is sitting is a fact *about him*. Consequently, the restricted-domain view fails to identify and reduce the existence of sitting Joe Biden with and to a REDO's having the property of containing him.

To sum up: The view's reductive explanatory strategy does not work because it wrongly assumes that the existence of an object X can be *separated* from X and identified with and reduced to a REDO's having the property of containing X and being [X]ish.[46] The existence of an object X cannot be separated from X because any object that exists exists as an object that has accidental properties.[47] And its having accidental properties cannot be identified with and reduced to a REDO's having properties. In other words, there is an intrinsic connection between an object X's having accidental properties and the existence of that object. Consequently, any explanation for what it is for an object X to exist must be able to account for this connection.

5.2.6 What Kind of Entity Is a REDO?

A sixth and last problem with the restricted-domain view is that it is unable to determine what kind of entity a REDO is.[48] The notion of a REDO—a 'field of

45 Cf. ibid., 133–137.
46 Cf. ibid., 135.
47 Gabriel accepts this claim by rejecting the idea of 'bare existence' or 'bare particulars' and committing himself to what he calls an 'ontological descriptivism' according to which objects exist always as objects which are this or that and so-and-so, cf. Gabriel, *Fields of Sense*, 12–13, 61, 95–96, 131; *Sinn und Existenz*, 139, 224–225.
48 My argumentative strategy in this section is inspired by Vallicella and follows him with regard to some distinctions but also significantly differs from his way of arguing, see Vallicella, *A Paradigm Theory of Existence*, 137–146.

sense' in Gabriel's terminology—plays a crucial role in the view's attempt to explain what it is for an object X to exist. Consequently, the restricted-domain view should be able to explain what kind of entity a REDO is. For example, is a restricted domain of objects just a collection of objects, or is it an entity which exists in its own right?

Whatever the restricted-domain view suggests that a REDO is, a REDO is not such that its having the property of containing an object X and being [X]ish could explain what it is for X to exist. In other words, there are no entities called 'restricted domains of objects' or 'fields of sense' which are such that their having the property of containing an object X and being [X]ish could explain what it is for X to exist.[49] No REDO is such that the existence of X can be identified with and reduced to that REDO's having the property of containing X and being [X]ish. To see why, let us consider systematically what kind of entity a REDO could be.

A first possibility which comes to mind when someone uses the term 'domain of objects' is that they are referring to a *collection of objects*.[50] A characteristic of a collection of objects is that it is nothing more than the sum of its parts or members, i.e., it is nothing more than the objects it contains.[51] For example, a collection of objects is the set of all the philosophy books in my office and the set of all the fish swimming in a given fish tank. The set of all the philosophy books in my office is nothing more than the books it contains because it is not itself a book which one could read or an object which one could discover in my office in addition to the books which are in it. In the same way, the set of all the fish swimming in a given fish tank is nothing more than the fish it contains because it is not itself a fish which one could catch or any other object which one could find in the fish tank in addition to the fish which are in it.

Now, a REDO cannot be a collection of objects. The core idea of the restricted-domain view is that the existence of an object X can be identified with and reduced to a REDO's having the property of containing X and being [X]ish. For X to exist is to be contained in a REDO or to appear in a REDO.[52] But this reductive explanation of what it is for X to exist works only if a REDO exists and a collection of objects does not exist. Consequently, a REDO must be a different kind of entity.

Against this argument one might object that it dismisses too quickly the possibility that a REDO is a collection of objects because collections of objects do

49 For this formulation of the problem, I am indebted to ibid., 137.
50 Cf. ibid., 138.
51 Cf. ibid., 137.
52 Cf. Gabriel, *Fields of Sense*, 44, 65, 140, 142, 166, 188–189, 192.

exist. It might be true that a collection of objects does not exist in itself, i.e., as an object *in addition to* the objects which are its members or parts. Nevertheless, a collection of objects exists in the sense that its existence *coincides* with the existence of the objects which are its parts or members. If the objects which are the members or parts of the collection exist, then the collection of objects which encompasses these objects exists as well.[53] Thus, a collection of objects exists but not in addition to the objects which are its members or parts. Rather, a collection of objects exists as the sum of all the objects which are its members or parts.

Whatever one may think about this way to account for the existence of collections of objects, it is of no help for Gabriel's view. First, Gabriel rejects the idea that restricted domains of objects—fields of sense in his terminology—are collections, sets, or mereological sums.[54] Second, if a collection of objects does not exist in addition to the objects which are its members or parts and exists only in the sense that its existence coincides with the existence of the objects which are its parts or members, it follows that its existence is *ontologically dependent* on the existence of the objects which are its members or parts. In other words, in this case, a collection of objects exists in virtue of the existence of its members or parts. For example, the set of all the philosophy books in my office exists in virtue of the philosophy books which exist in my office. It is the existence of the philosophy books which explains the existence of the set. Consequently, the explanation of what it is for any of my philosophy books to exist cannot be that the set or collection of the philosophy books in my office has the property of containing that book. Such an explanation is viciously circular because the existence of the collection of the philosophy books in my office is ontologically dependent on the existence of its members. Thus, what is supposed to explain what it is for an object X to exist (e.g., a collection of objects' property of containing X and being [X]ish), in fact presupposes the existence of X.[55]

To sum up: A REDO cannot be a collection of objects because a collection of objects is not such that its having the property of containing an object X and being [X]ish could explain what the existence of an object X consists in. Either a collection of objects does not exist so that the existence of X cannot be identified with and reduced to the collection's having the property of containing X and being [X]ish, or a collection of objects exists but its existence is ontologically dependent on the existence of X with the consequence that the view's explana-

---

53 Cf. Vallicella, *A Paradigm Theory of Existence*, 137.

54 Cf. Gabriel, *Fields of Sense*, 12, 94–95, 160, 229; "Repliken auf Diehl/Rosefeldt, Hübner, Rödl und Stekeler-Weithofer," 189–192.

55 Cf. Vallicella, *A Paradigm Theory of Existence*, 138–139.

tion of what it is for X to exist becomes viciously circular. Thus, the restricted-domain view must discard the possibility that restricted domains of objects are collections, sets, or mereological sums of objects.

This brings us to a second possibility. To evade the above-mentioned problems, the restricted-domain view could argue that a REDO is not a collection of objects but an entity which meets two requirements:[56] First, a REDO is an entity which exists. Second, a REDO is an entity whose existence is *ontologically independent* from the objects which it contains or which appear in it.

These two requirements are met by some restricted domains of objects or fields of sense which Gabriel and others use to illustrate what it is for an object X to exist according to their view: The existence of a particular human being called 'Pierre' consists in a particular café's having the property of containing Pierre and being [Pierre]ish and the existence of the café consists in Paris's having the property of containing the café and being [café]ish.[57] The existence of a particular pig—let us continue to call her 'Piggy'—consists in a particular farm's having the property of containing Piggy and being [Piggy]ish.[58] The existence of the coffee in Markus Gabriel's mug consists in his mug's having the property of containing coffee and being [coffee]ish, the existence of the mug consists his room's having the property of containing the mug and being [mug]ish, and the existence of Gabriel's computer at time t consists in his living room's having the property of containing his computer and being [computer]ish at time t.[59]

In each of these cases, the relevant REDO meets the two requirements: The existence of the café is ontologically independent from the existence of Pierre, the existence of Paris is ontologically independent from the existence of the café, the existence of the farm is ontologically independent from the existence of Piggy, the existence of Gabriel's mug is ontologically independent from the existence of coffee, the existence of the room is ontologically independent from the existence of the mug and the existence of Gabriel's living room at time t is ontologically independent from the existence of his computer at time t. Moreover, the living room, the mug, the farm, Paris, and the café are all entities which are supposed to exist. They must exist because it would be absurd to claim that an object X exists in virtue of something Y's having the property of containing X if Y does not exist.

---

56 Cf. ibid., 138, 144.
57 Cf. ibid., 146.
58 Cf. Gabriel, *Fields of Sense*, 60.
59 Cf. ibid., 160, 168.

This brings us to the following problem: If a REDO is an entity which exists and whose existence is *ontologically independent* from an object X it contains or which appears in it, then the property of containing X and being [X]ish is merely accidental to a REDO.[60] A REDO can cease to contain X and to be [X]ish without therefore ceasing to exist. For example, Gabriel can drink the coffee in his mug without it being the case that his mug ceases to exist. But if the property of containing X and being [X]ish is accidental to a REDO, the decisive question is what happens to X if a REDO loses the property of containing X?

Gabriel's view is caught in the following dilemma: Either a REDO's having the property of containing X and being [X]ish is constitutive of the existence of X or it is not. If a REDO's having the property of containing X and being [X]ish is *constitutive* of the existence of X, the existence of X is ontologically dependent on the existence of the REDO. This is absurd. The existence of Pierre is not ontologically dependent on the existence of the café in Paris, the existence of Piggy is not ontologically dependent on the farm she lives on etc. In other words, if the existence of X is ontologically dependent on the existence of a REDO, X is not an object which exists in its own right but an object which exists as an accident or modification of a REDO (see 5.2.4).

Alternatively, a REDO's having the property of containing X and being [X]ish is *not constitutive* of the existence of X. If a REDO's having the property of containing X and being [X]ish is not constitutive of the existence of X, the existence of X is ontologically independent from the existence of the REDO. In this case, the ontological status of X as an object which exists in its own right is preserved. However, if a REDO's having the property of containing X and being [X]ish is not constitutive of the existence of X and the existence of X is ontologically independent from the existence of the REDO, the restricted-domain view fails to explain what it is for X to exist. In other words, the view's attempt to explain what it is for X to exist by identifying and reducing the existence of X with and to a REDO's having the property of containing X and being [X]ish fails.

This result is fatal for the restricted-domain view because the two possibilities considered are mutually exclusive and exhaustive. Either a REDO is a collection of objects or it is not. If it is, it either exists itself or not. In the latter case, the view's explanation of what it is for an object X to exist does not work because it would be absurd to claim that the existence of X consists in something non-existing's property of containing X and being [X]ish. In the former case, the view's explanation of what it is for an object X to exist is viciously circular because a

60 Cf. Vallicella, *A Paradigm Theory of Existence*, 144.

REDO as a collection of objects exists but its existence is ontologically dependent on the existence of X as its member or part.

If a REDO is not a collection of objects whose existence is dependent on its members or parts but an entity whose existence is ontologically independent from the objects it contains or which appear in it, the restricted-domain view faces another dilemma. Either its explanation of what it is for an object X to exist undermines the ontological status of X as an object which exists in its own right, or the view's explanation does not work because a REDO's having the property of containing X and being [X]ish is not constitutive of the existence of X. To sum up: There are no entities called 'restricted domains of objects' which are such that their having the property of containing an object X and being [X]ish could explain what it is for an object X to exist.

Part II: The Contribution of Thomas Aquinas's View to the Debate

The first part of my book offered a systematic reconstruction of the contemporary debate about the nature question. I presented five views on what it is for an object to exist which have been entertained in the contemporary discussion. The first view on what it is for an object to exist employs a *non-reductive* explanatory strategy (see 1). According to this explanatory strategy, the existence of an object X consists in X's having a property *sui generis*, namely, the property of existing. Thus, for X to exist is to have a property *sui generis*, namely, the property of existing. As a property *sui generis* existence is a primitive and undefinable property which cannot be identified with and reduced to any other property that X has.

What distinguishes the other four answers to the nature question from the one which is given by the non-reductive-object view is that they employ a *reductive* explanatory strategy to explain what it is for an object to exist. What unites these four views is the idea that the existence of an object X can be identified with and reduced to something's having a property different from existence.

What distinguishes the four views from each other is what they attempt to identify the existence of an object X with and reduce to. In Chapter 2, I reconstructed the view that the existence of an object X consists in that *object X's* having a property, namely, the property of being numerically identical with itself.

In contrast to this reductive-object view, the remaining three views claim that the existence of an object X consists in *something else Y's* having a property. Chapter 3 addressed the view that the existence of an object X consists in a property's having the property of being instantiated. In Chapter 4, I engaged with the view that the existence of an object X consists in an *unrestricted* domain of objects'—a UDO's—property of containing X and being [X]ish. Finally, in the last Chapter of the first part, I dealt with the view that the existence of an object X consists in a *restricted* domain of objects'— a REDO's—property of containing X and being [X]ish.

The upshot of this systematic reconstruction of the contemporary debate about the nature question is that none of the considered views can explain satisfyingly what it is for an object to exist. Their attempts to explain what it is for an object to exist are riddled with problems. In other words, I have shown that there are good reasons to reject the answers to the nature question these views are giving. So, my first thesis is substantiated. This leads us to the second, constructive thesis.

In the two remaining chapters of this second part of my book, I will defend the claim that Thomas Aquinas has answered the nature question in a better way. Chapter 6 reconstructs in three steps how Aquinas explains what it is for an object to exist, i.e., a substance in his terminology. In Chapter 7, I explain in detail why Aquinas's view can solve the problems that the views discussed in the first part cannot.

I will conclude that Aquinas's view not only solves the problem of what it is for an object to exist, but also contributes a new perspective and vantage point which allows us to see why the contemporary debate about the nature question has reached an impasse, i.e., why the contemporary views I discussed are unable to formulate a satisfactory answer to the question of what it is for an object to exist.

6 Aquinas's View on What It Is for a Substance to Exist

6.1 For a Substance to Exist Is to Have *Esse* by Means of Its Essence or Form

What is Aquinas's answer to the nature question? What is his view on what it is for an object X to exist, i.e., a substance in his terminology? In this section, I will reconstruct a first preliminary answer to this question. To understand Aquinas's view on what it is for a substance to exist, we need to familiarize ourselves, at least to a certain extent, with the terminology and the general metaphysical framework within which he formulates his view. In particular, his views on *ens*, *esse*, form, and essence play an important role in his explanation of what it is for a substance to exist.[1]

Let us begin with Aquinas's answer to the inventory question. What is there or what exists outside the human mind, i.e., independently of the human mind, according to Aquinas? That which is or exists outside the human mind are *entia* which are actual.[2] An *ens* which is actual is an *ens* which is the source of some kind of action, activity, or operation.[3] In other words, what accounts for the ex-

1 I leave the terms '*ens*' and '*esse*' untranslated because this makes it easier to follow the train of thought in the latin quotes from Aquinas I give. Furthermore, any translation—e.g., translating '*esse*' as '*existence*'—might have unwelcome connotations which would make it more difficult to understand what Aquinas's view on the matter is.

2 Cf. *DSS*, cap. 8: "Omne autem quod est esse habet; est igitur in quocumque praeter primum et ipsum esse tanquam actus, et substantia rei habens esse tanquam potentia receptiva huius actus quod est esse."; *SCG*, I, cap. 22: "Omnis res est per hoc quod habet esse."; ibid.: "Nulla igitur res cuius essentia non est suum esse, est per essentiam suam, sed participatione alicuius, scilicet ipsius esse."

3 Cf. *SCG*, I, cap. 28: "Nihil agit nisi secundum quod est in actu. Actio igitur consequitur modum actus in agente."; *QDP*, q. 3, a.1, corp: "Ad cuius evidentiam sciendum est, quod omne agens agit, secundum quod est actu; unde oportet quod per illum modum actio alicui agenti attribuatur quo convenit ei esse in actu."; *SCG*, II, cap. 7: "Sicut potentia passiva sequitur ens in potentia, ita potentia activa sequitur ens in actu: unumquodque enim ex hoc agit quod est actu, patitur vero ex hoc quod est potential."; *SCG*, III, cap. 69: "Si agere sequitur ad esse in actu, inconveniens est quod actus perfectior actione destituatur."; *SCG*, III, cap. 97: "Cum enim unumquodque agat secundum quod est actu, quae enim sunt in potentia, secundum quod huiusmodi, inveniuntur actionis expertia; est autem unumquodque ens actu per formam: oportet quod operatio rei sequatur formam ipsius."; *DSS*, cap. 17: "Primo namque penitus irrationale est ut malorum ponatur esse aliquod primum principium quasi contrarium summo bono. Nihil enim potest esse activum nisi in quantum est ens actu, quia unumquodque tale alterum agit quale ipsum est; rursumque ex

https://doi.org/10.1515/9783110979879-009

istence of an *ens* outside of the human mind is its actuality—its *esse* in Aquinas's terminology—which manifests itself to us through some kind of action, activity, or operation of the relevant *ens*.

In Aquinas's view, what exists outside the human mind is only one *ens* which is subsistent *esse* itself, i.e., an *ens* whose essence is *esse* and which is therefore pure actuality, and a multiplicity of *entia* which have *esse* and participate in *esse* and are therefore metaphysical composites of potentiality and actuality.[4] In this inquiry, I am not concerned with Aquinas's view on that which *is* subsistent *esse* itself and therefore pure actuality. I am only interested in that which *has esse* and which therefore is a metaphysical composite of actuality

---

hoc aliquid agitur quod actu fit." On action or activity as Aquinas's criterion for the actuality of an *ens*, see also Clarke, *Explorations*, 45–64; *The One and the Many: A Contemporary Thomistic Metaphysics* (Notre Dame, Indiana: University of Notre Dame Press, 2001), 31–36.

4 Cf. *SCG*, I, cap. 22: "Nulla igitur res cuius essentia non est suum esse, est per essentiam suam, sed participatione alicuius, scilicet ipsius esse."; *SCG*, II, cap. 15: "Deus autem est ens per essentiam suam: quia est ipsum esse. Omne autem aliud ens est ens per participationem: quia ens quod sit suum esse non potest esse nisi unum ut in primo ostensum est."; *SCG*, II, cap. 52: "Illud ergo quod est esse subsistens, non potest esse nisi unum tantum. Ostensum est autem quod Deus est suum esse subsistens. Nihil igitur aliud praeter ipsum potest esse suum esse. Oportet igitur in omni substantia quae est praeter ipsum, esse aliud ipsam substantiam et esse eius. [...] Ipsum esse competit primo agenti secundum propriam naturam: esse enim Dei est eius substantia ut supra ostensum est. Quod autem competit alicui secundum propriam naturam suam, non convenit aliis nisi per modum participationis: sicut calor aliis corporibus ab igne. Ipsum igitur esse competit omnibus aliis a primo agente per participationem quandam. Quod autem competit alicui per participationem, non est substantia eius. Impossibile est igitur quod substantia alterius entis praeter agens primum sit ipsum esse."; *SCG*, II, cap. 53: "Omne participans aliquid comparatur ad ipsum quod participatur ut potentia ad actum: per id enim quod participatur fit participans actu tale. Ostensum autem est supra quod solus Deus est; essentialiter ens, omnia autem alia participant ipsum esse. Comparatur igitur substantia omnis creata ad suum esse sicut potentia ad actum."; *QQ*, XII, a.1, corp.: "Sciendum ergo quod unumquodque quod est in potentia et in actu, fit actu per hoc quod participat actum superiorem; per hoc autem aliquid maxime fit actu, quod participat per similitudinem primum et purum actum; primus autem actus est esse subsistens per se; unde completionem unumquodque recipit per hoc quod participat esse. Unde esse est completivum omnis formae, quia per hoc completur quod habet esse, et habet esse cum est actu; et sic nulla forma est nisi per esse."; *QDA*, q. 6, corp: "Et ita in formis per se subsistentibus invenitur et potentia et actus, in quantum ipsum esse est actus formae subsistentis, quae non est suum esse. Si autem aliqua res sit quae sit suum esse—quod proprium Dei est—, non est ibi potentia et actus, sed actus purus. Et hinc est quod Boethius dicit in libro De hebdomadibus, quod in aliis quae sunt post Deum differt esse et quod est, vel, sicut quidam dicunt, quod est et quo est, nam ipsum esse est quo aliquid est, sicut cursus est quo aliquis currit. Cum igitur anima sit quaedam forma per se subsistere potens, est in ea compositio actus et potentiae, scilicet esse et quod est, non autem compositio materiae et formae."

and potentiality.[5] According to Aquinas, that which has *esse* outside the human mind are *entia* which belong to one of the ten Aristotelian categories.[6]

Besides *entia* which exist outside and independently of the human mind, Aquinas also recognizes *entia* like fictional entities (e. g., a goat-stag or a phoenix) or privations (e. g., blindness) which only exist *in* the human mind and whose existence therefore depends on the human mind.[7] The latter *entia* are called '*entia*' by Aquinas because they too have *esse*. But the *esse* they have is not the *esse* which *entia* outside the human mind have.

Entia which exist only in the human mind do not have *esse* in the sense of actuality like *entia* which exist outside the human mind. *Entia* which exist only in the human mind have *esse* only in the sense of the truth of a proposition. In other words, the only *esse* they have is the *esse* which is attributed to them in the process in which the human mind forms an affirmative proposition about them.[8]

---

5 I am not concerned with Aquinas's view on that which is subsistent *esse* itself because in this book I deal with the nature or weak ground question and not with the strong ground question. According to Aquinas, a reference to subsistent *esse* itself is necessary to explain why something has *esse* and exists. Now, in this inquiry, I simply assume that something has *esse* and I am just interested to know how this 'having' could be explained. In other words, this book is not looking for a kind of causal explanation which could answer the question why there is something rather than nothing. Rather, I am looking for an answer to the question of what it is for that which exists to exist. Furthermore, the nature question does not arise for an *ens* which is subsistent *esse* itself because such an *ens* does not have *esse* but is *esse*. For such an *ens* to exist is just to be what it is, namely, *esse*.

6 Cf. *DEE*, cap.1: "Sciendum est igitur quod, sicut in V Metaphysicae Philosophus dicit, ens per se dupliciter dicitur: uno modo quod dividitur per decem genera, alio modo quod significat propositionum veritatem."; *QQ*, IX, q. 2, a. 2, corp.: "Alio mode esse dicitur actus entis in quantum est ens, id est quo denominatur aliquid ens actu in rerum natura; et sic esse non attribuitur nisi rebus ipsis quae in decem generibus continentur, unde ens a tali esse dictum per decem genera dividitur."

7 In Aquinas's view, blindness is an *ens* which exists only in the human mind because blindness is not a form but the privation of a form, namely, the form of sight. Any form of a substance is an *ens* which gives *esse* to a substance, i. e., a form of a substance is an *ens* by which the substance is actual in a certain respect. But blindness is not a form and does not give *esse* to a substance. On the contrary, if a substance is blind, it lacks a form—the form of sight—and consequently the *esse* it should have. So, in Aquinas's view, privations like blindness are *entia* which the mind needs to produce to be able to express that something outside the human mind lacks a form and consequently the *esse* it should have. I am thankful to Jason T. Eberl for making me aware of the need to explain why an *ens* like blindness is mind-dependent.

8 Cf. *DEE*, cap.1: "Horum autem differentia est quia secundo modo potest dici ens omne illud de quo affirmativa propositio formari potest, etiam si illud in re nihil ponat; per quem modum privationes et negationes entia dicuntur: dicimus enim quod affirmatio est opposita negationi, et quod caecitas est in oculo. Sed primo modo non potest dici ens nisi quod aliquid in re ponit; unde primo modo caecitas et huiusmodi non sunt entia."; *QQ*, IX, q. 2, a. 2, corp.: "Dicendum

Such entities 'exist' only in the qualified sense of existing in the mind. They do not exist in the sense of having actuality.

In this inquiry, I do not deal with the question of what it is for an entity to exist which exists only in the human mind. Rather, I am interested in the question of what it is for an entity to exist which exists independently of any human mind. Therefore, this chapter is limited to an exposition of Aquinas's view on what the existence of an *ens*, which has *esse* in the sense of actuality, consists in.[9]

According to Aquinas, all *entia* which belong to one of the ten Aristotelian categories have *esse* in the sense of actuality.[10] However, such *entia* do not have *esse* in the same way.[11] What properly and truly exists outside the human mind are *entia* which belong to the category of substance. All other *entia* are *entia* only insofar as they exist with and in substances.[12]

---

quod esse dupliciter dicitur, ut patet per Philosophum in V Metaphysicae et in quadam glossa Origenis super principium Iohannis. Uno mode secundum quod est copula verbalis significans compositionem cuiuslibet enunciationis quam anima facit, unde hoc esse non est aliquid in rerum natura, sed tantum in actu animae componentis et dividentis; et sic esse attribuitur omni ei de quo potest propositio formari, sive sit ens sive privatio entis: dicimus enim caecitatem esse."; *QQ*, IX, q. 2, a. 2, ad 4: "Ad quartum dicendum quod obiectio illa procedit de esse quod in actu animae consistit, secundum quod etiam ea quae sunt tantum secundum animam esse dicuntur, cum per 'an est' etiam de caecitate quaeri possit."

9 For Aquinas's views on what it is for *entia* to exist which only exist in the human mind, see, for example, Gabriele Galluzzo, "Aquinas on Mental Being," *Quaestio* 2010, no. 10 (2010); Gyula Klima, "The Changing Role of 'Entia Rationis' in Medieval Semantics and Ontology: A Comparative Study with a Reconstruction," *Synthese* 96, no. 1 (1993); Gloria Wasserman, "Thomas Aquinas on Truths About Nonbeings," *Proceedings of the American Catholic Philosophical Association* 80 (2006).

10 Cf. *QQ*, IX, q. 2, a. 2, corp.: "Alio mode esse dicitur actus entis in quantum est ens, id est quo denominatur aliquid ens actu in rerum natura; et sic esse non attribuitur nisi rebus ipsis quae in decem generibus continentur, unde ens a tali esse dictum per decem genera dividitur."

11 Cf. *QQ*, II, q. 2, a.1, corp.: "Et ideo Commentator dicit in V Metaphysicae quod ista propositio: Sortes est, est de accidentali praedicato, secundum quod importat entitatem rei vel veritatem propositionis, sedverum est quod hoc nomen ens, secundum quod importat rem cui competit huiusmodi esse, sic significat essentiam rei, et dividitur per decem genera. Non tamen univoce, quia non eadem ratione competit omnibus esse, sed substantiae quidem per se, aliis autem aliter."

12 Cf. *DEE*, cap. 1: "Sed quia ens absolute et primo dicitur de substantiis, et per posterius et quasi secundum quid de accidentibus, inde est quod essentia proprie et vere est in substantiis, sed in accidentibus est quodammodo et secundum quid."; *QQ*, IX, q. 2, a. 2, corp.: "Sed hoc esse attribuitur alicui rei dupliciter. Uno modo, sicut ei quod proprie et vere habet esse vel est; et sic attribuitur soli substantiae per se subsistenti, unde quod vere est dicitur substancia in I Physicorum. Omnibus vero quae non per se subsistunt sed in alio et cum alio, sive sint accidentia sive formae substantiales aut quaelibet partes, non habent esse ita quod ipsa vere sint, sed attribuitur eis esse alio modo, id est ut quo aliquid est, sicut albedo dicitur esse, non quia ipsa in esse

The most fundamental categorical distinction is between *entia* which belong to the category of substance and *entia* which belong to one of the other nine categories of accidents (e.g., quality, quantity, relation etc.). *Entia* which belong to the category of substance have in common that they have a quiddity or essence to which it belongs to have *esse* not in another as a subject. In contrast, *entia* which belong to one of the nine categories of accidents have in common that they have a quiddity or essence to which it belongs to have *esse* in another as a subject.[13] In other words, an *ens* which belongs to the category of substance has *esse per se*, i.e., is a subsistent thing (*hoc aliquid*), while an *ens* which belongs to one of the nine Aristotelian categories of accidents does not have *esse per se* but has *esse* only in another as a subject and is therefore not a subsistent thing.[14] Thus, in Aquinas's view, the most important categorical division is the division between *entia* which have *esse per se*, i.e. substances which are subsistent things, and *entia* which have *esse* in another as a subject, i.e., accidents which are non-subsistent things.[15]

---

subsistat, sed quia ea aliquid habet esse album. Esse ergo proprie et vere non attribuitur nisi rei per se subsistenti."

13 Cf. *ST*, III, q. 77, a. 1, ad. 2: "Ad secundum dicendum quod, cum ens non sit genus, hoc ipsum quod est esse, non potest esse essentia vel substantiae vel accidentis. Non ergo definitio substantiae est ens per se sine subiecto, nec definitio accidentis ens in subiecto sed quidditati seu essentiae substantiae competit habere esse non in subiecto; quidditati autem sive essentiae accidentis competit habere esse in subiecto." For the English paraphrase of Aquinas's Latin, I am relying here on Christopher M. Brown, *Aquinas and the Ship of Theseus: Solving Puzzles About Material Objects* (London: Continuum, 2005), 50–51.

14 Cf.: *QQ*, IX, q. 2, a. 2, corp.: "Sed hoc esse attribuitur alicui rei dupliciter. Uno modo, sicut ei quod proprie et vere habet esse vel est; et sic attribuitur soli substantiae per se subsistenti, unde quod vere est dicitur substancia in I Physicorum. Omnibus vero quae non per se subsistunt sed in alio et cum alio, sive sint accidentia sive formae substantiales aut quaelibet partes, non habent esse ita quod ipsa vere sint, sed attribuitur eis esse alio modo, id est ut quo aliquid est, sicut albedo dicitur esse, non quia ipsa in esse subsistat, sed quia ea aliquid habet esse album. Esse ergo proprie et vere non attribuitur nisi rei per se subsistenti."; *QDP*, q.3, a. 8, corp.: "Forma enim naturalis non dicitur univoce esse cum re generata. Res enim naturalis generata dicitur esse per se et proprie, quasi habens esse, et in suo esse subsistens; forma autem non sic esse dicitur, cum non subsistat, nec per se esse habeat; sed dicitur esse vel ens, quia ea aliquid est; sicut et accidentia dicuntur entia, quia substantia eis est vel qualis vel quanta, non quod eis sit simpliciter sicut per formam substantialem: unde accidentia magis proprie dicuntur entis, quam entia, ut patet in Metaphys."; *ST*, I, q. 75, a.2, ad 1: "Ad primum ergo dicendum quod hoc aliquid potest accipi dupliciter, uno modo, pro quocumque subsistente, alio modo, pro subsistente completo in natura alicuius speciei. Primo modo, excludit inhaerentiam accidentis et formae materialis, secundo modo, excludit etiam imperfectionem partis."

15 I am thankful to John Peck for helping me to see that a previous formulation of this distinction needed to be improved.

It is important to note that not every subsistent thing (*hoc aliquid*) is a substance for Aquinas. For example, in Aquinas's view, both a hand and the soul of a particular human being are subsistent things but not substances.[16] The difference between subsistent things which are substances and subsistent things which are not substances can be explained through a difference in the way they have *esse*. From what we have seen so far, it is clear that subsistent things have in common that they have *esse per se*. In virtue of having *esse per se*, they are distinguished from accidental forms and substantial forms of non-human material substances which do not subsist but exist only as *entia* which have *esse* in another as a subject.[17]

What distinguishes substances from subsistent things like a human hand and a human soul is that the former have *esse per se* and *in se* while the latter have *esse per se* but do not have *esse in se*. They subsist only in another—a substance—as a part of it.[18] Thus, having *esse per se* is a necessary but not a sufficient condition for being a substance.

Aquinas's point can be explained in the following way: A material substance like a human being is a subsistent thing which subsists in itself (*in se*), in contrast to a human hand or a human soul which are only subsistent things which subsist in something else—a human substance—as an integral or metaphysical part of it.[19] Aquinas distinguishes between a material substance as a *whole*

---

16 *ST* I, q. 75, a. 2, ad. 1: "Ad primum ergo dicendum quod hoc aliquid potest accipi dupliciter, uno modo, pro quocumque subsistente, alio modo, pro subsistente completo in natura alicuius speciei. Primo modo, excludit inhaerentiam accidentis et formae materialis, secundo modo, excludit etiam imperfectionem partis. Unde manus posset dici hoc aliquid primo modo, sed non secundo modo. Sic igitur, cum anima humana sit pars speciei humanae, potest dici hoc aliquid primo modo, quasi subsistens, sed non secundo modo, sic enim compositum ex anima et corpore dicitur hoc aliquid."

17 Cf. ibid.

18 Cf. *QDUVI*, a. 2, corp.: "Est autem substantiae proprium ut per se et in se subsistat; accidentis autem est in alio esse. Et ideo illa nomina quae pertinent ad individuationem substantiae, in illis solum locum habent quae per se et in se subsistunt. Et propter hoc etiam de partibus substantiarum non dicuntur, quia non sunt in seipsis sed in toto, quamvis non sint in subiecto." The idea to characterize substances in this way is inspired by Stump, *Aquinas*, 484–485 n.47.

19 The notion 'metaphysical part' is not used by Aquinas but introduced by contemporary scholars to explain how he—or any hylomorphist in general—distinguishes form and matter and a substance's matter-form composition from other quantitative parts a substance might have and a corresponding mereological composition of a material substance, cf. Robert Pasnau, *Metaphysical Themes, 1274–1671* (Oxford: Clarendon Press, 2011), 6–11; Stump, *Aquinas*, 35, 42, 52–53, 113, 414, 481 n.3, 481–482 n.5; Brown, *Aquinas and the Ship of Theseus*, 53, 71–75, 87–94.

which subsists *per se* and *in se* and an integral or metaphysical *part* of it which subsists not in itself (*in se*) but in the substance of which it is a part.[20]

Integral parts of material substances are, for example, their extremities (e. g., legs, hands, feet, heads, wings etc.) or organs (e. g., heart, liver, lungs etc.). Aquinas's example for a metaphysical part which subsists in a material substance is the substantial form of a human substance, i. e., the human soul.[21] The decisive difference between the substantial form of a human substance and the substantial form of a non-human material substance is that the former can have *esse* in separation from the matter it informs or configures.[22] Thus, it is important to note that, in Aquinas's view, the substantial form of a human being—the human soul —is an exception as a metaphysical part of material substance. The human soul is a metaphysical part of human beings which subsists, i. e., which has *esse per se*. All other substantial forms of material substances are metaphysical parts of them which do not subsist but have *esse* only in another as a subject.

Consequently, subsistent things which have *esse* not *in se* but in another as an integral or metaphysical part of it have only an incomplete essence or nature, in contrast to subsistent things which have *esse in se*.[23] A substance is a subsis-

---

20 Cf. *QDUVI*, a. 2, corp.: "Et ideo illa nomina quae pertinent ad individuationem substantiae, in illis solum locum habent quae per se et in se subsistunt. Et propter hoc etiam de partibus substantiarum non dicuntur, quia non sunt in seipsis sed in toto, quamvis non sint in subiecto."
21 Cf. *ST*, I, q. 75, a. 2, ad. 1: "Ad primum ergo dicendum quod hoc aliquid potest accipi dupliciter, uno modo, pro quocumque subsistente, alio modo, pro subsistente completo in natura alicuius speciei. Primo modo, excludit inhaerentiam accidentis et formae materialis, secundo modo, excludit etiam imperfectionem partis. Unde manus posset dici hoc aliquid primo modo, sed non secundo modo. Sic igitur, cum anima humana sit pars speciei humanae, potest dici hoc aliquid primo modo, quasi subsistens, sed non secundo modo, sic enim compositum ex anima et corpore dicitur hoc aliquid." This does not mean that all metaphysical parts of material substances subsist. For example, in Aquinas's view, the substantial form of a non-human material substance is a non-subsistent metaphysical part of it as well as the designated matter of any material substance which does not subsist independently from the substantial form which informs it. I owe my thanks to Jason T. Eberl for making me aware that this issue needed to be clarified.
22 Cf. *QDA*, q. 6, corp.: "Et licet materia non pertingat ad esse nisi per formam, forma tamen, in quantum est forma, non indiget materia ad suum esse, cum ipsam formam consequatur esse, sed indiget materia cum sit talis forma quae per se non subsistat. Nihil igitur prohibet esse aliquam formam a materia separatam quae habeat esse. [...] Cum igitur anima sit quaedam forma per se subsistere potens, est in ea compositio actus et potentiae, scilicet esse et quod est, non autem compositio materiae et formae."
23 Cf. *ST*, I, q. 75, a. 2, ad. 1: "Ad primum ergo dicendum quod hoc aliquid potest accipi dupliciter, uno modo, pro quocumque subsistente, alio modo, pro subsistente completo in natura alicuius specie." It is this feature which integral and metaphysical parts share with accidental forms whose essences or natures are also incomplete because their quiddity or essence cannot

tent thing with a complete essence or nature because it has a quiddity or essence whose definition does not include a reference to something extrinsic to it.[24] In contrast, a human hand or a human soul is a subsistent thing with an incomplete essence or nature because it has a quiddity or essence whose definition does include a reference to something extrinsic to it, namely, a reference to the whole of which it is an integral or metaphysical part.[25] For example, a heart or leg is what it is in virtue of the material substance of which it is an integral part and a human substantial form—a human soul—is what it is in virtue of the human substance of which it is a metaphysical part.[26]

To sum up: In Aquinas's view, what properly and truly exists outside the human mind are substances, i.e., *entia* which have *esse per se* and *in se*.[27] Every other *ens* that has *esse* has *esse* in a different way and exists therefore only in a qualified way.[28]

So far, I have provided a sketch of Aquinas's answer to the inventory question, i.e., the question of what there is or exists. Let us now turn to Aquinas's answer to the nature question. What is it for an *ens*, i.e., a substance, *to exist?* As we have already seen above, in Aquinas's view, a substance is an *ens* which exists outside the human mind in virtue of having *esse* in the sense of ac-

---

be defined without a reference to something which is extrinsic to it. In the case of accidental forms, their quiddity or essence cannot be defined without a reference to a subject in which they inhere, cf. *DEE*, cap. 6: "Et quia, ut dictum est, essentia est id quod per diffinitionem significatur, oportet ut eo modo habeant essentiam quo habent diffinitionem. Diffinitionem autem habent incompletam, quia non possunt diffiniri nisi ponatur subiectum in eorum diffinitione." I am thankful to James Dominic Rooney for making me aware of the need to clarify this point.

24 Cf. Brown, *Aquinas and the Ship of Theseus*, 55.

25 Cf. ibid. I am indebted to James Dominic Rooney who helped me to improve the formulation of this difference.

26 Cf. ibid., 55–56; Stump, *Aquinas*, 42.

27 Cf. *ST*, I, q. 75, a. 2, ad. 2: "Sed per se existens quandoque potest dici aliquid si non sit inhaerens ut accidens vel ut forma materialis, etiam si sit pars. Sed proprie et per se subsistens dicitur quod neque est praedicto modo inhaerens, neque est pars."; *QQ*, IX, a. 2, corp.: "Uno modo, sicut ei quod proprie et vere habet esse vel est; et sic attribuitur soli substantiae per se subsistenti, unde quod vere est dicitur substancia in I Physicorum. [...] Esse ergo proprie et vere non attribuitur nisi rei per se subsistenti."

28 Cf. *QDV*, q. 21, a. 4, ad 4: "Aliquid enim dicitur ens, quia in se subsistit; aliquid, quia est principium subsistendi, ut forma; aliquid, quia est dispositio subsistentis, ut qualitas; aliquid, quia est privatio dispositionis subsistentis, ut caecitas. Et ideo cum dicimus: essentia est ens; si procedatur sic: ergo est aliquo ens, vel se vel alio; processus non sequitur, quia non dicebatur hoc modo esse ens, sicut aliquid subsistens in esse suo est ens, sed sicut quo aliquid est. Unde non oportet quaerere quomodo ipsa essentia aliquo sit sed quomodo aliquid alterum sit per essentiam."

tuality.[29] In other words, the existence of a substance consists in a substance's having *esse*. But this is not all that Aquinas has to say about the matter. Rather, Aquinas has a sophisticated way to explain how this 'having' works, i.e., how a substance 'has' the *esse* it has. In what follows, I will argue that Aquinas's view on what it is for a substance to exist can be reconstructed in three stages.

As a first stage, Aquinas claims that for a substance to exist is to have *esse* by means of its essence or form. More generally speaking, a substance has *esse* by mediation of some form.[30] A substance has accidental *esse* by means of an accidental form and it has substantial *esse* by means of its substantial form.[31]

Let us begin with Aquinas's explanation of a substance's having accidental *esse*. According to Aquinas, an accidental form of a substance is that by means of which a substance exists as a substance which is actual in some accidental respect, for example, with respect to its quality or quantity.[32] By means of an ac-

29 Cf. *DSS*, cap. 8: "Omne autem quod est esse habet."; *SCG*, I, cap. 22: "Omnis res est per hoc quod habet esse."

30 Cf. *SCG*, II, cap. 73: "Sed unumquodque habet esse per suam formam."; *DEE*, cap. 1: "Sed essentia dicitur secundum quod per eam et in ea ens habet esse."; *QDV*, q. 27, a.1, ad. 3: "Ad tertium dicendum, quod esse naturale per creationem Deus facit in nobis nulla causa agente mediante, sed tamen mediante aliqua causa formali: forma enim naturalis principium est esse naturalis."; For a more extensive analysis of the mediating role of form, see Rudi A. Te Velde, *Participation and Substantiality in Thomas Aquinas* (Leiden; New York; Köln: E.J. Brill, 1995), 212–233.

31 Cf. *QQ*, IX, a.2., corp.: "Omnibus vero quae non per se subsistunt sed in alio et cum alio, sive sint accidentia sive formae substantiales aut quaelibet partes, non habent esse ita quod ipsa vere sint, sed attribuitur eis esse alio modo, id est ut quo aliquid est, sicut albedo dicitur esse, non quia ipsa in esse subsistat, sed quia ea aliquid habet esse album."; *ST*, III, q. 17, a. 2, corp.: "Natura enim significatur per modum formae, quae dicitur ens ex eo quod ea aliquid est, sicut albedine est aliquid album, et humanitate est aliquis homo."; *QDUVI*, a. 4, corp.: "Esse enim proprie et vere dicitur de supposito subsistente. Accidentia enim et formae non subsistentes dicuntur esse, in quantum eis aliquid subsistit; sicut albedo dicitur ens, in quantum ea est aliquid album. Considerandum est autem, quod aliquae formae sunt quibus est aliquid ens non simpliciter, sed secundum quid; sicut sunt omnes formae accidentales. Aliquae autem formae sunt quibus res subsistens simpliciter habet esse; quia videlicet constituunt esse substantiale rei subsistentis."; *DEE*, cap. 6: "Et hoc ideo est quia non habent esse per se absolutum a subiecto, sed sicut ex forma et materia relinquitur esse substantiale quando componuntur, ita ex accidente et subiecto relinquitur esse accidentale quando accidens subiecto advenit. [...] Et ideo accidens superveniens ex coniunctione sui cum eo cui advenit non causat illud esse in quo res subsistit, per quod res est ens per se; sed causat quoddam esse secundum sine quo res subsistens intelligi potest esse, sicut primum potest intelligi sine secundo."

32 Cf. *QDP*, q. 3, a. 8, corp.: "Forma enim naturalis non dicitur univoce esse cum re generata. Res enim naturalis generata dicitur esse per se et proprie, quasi habens esse, et in suo esse subsistens; forma autem non sic esse dicitur, cum non subsistat, nec per se esse habeat; sed dicitur esse vel ens, quia ea aliquid est; sicut et accidentia dicuntur entia, quia substantia eis est vel

cidental form (*quo est*), a substance (*quod est*) exists as a substance which is modified in an accidental respect.[33] Thus, in Aquinas's view, an accidental form of a substance is an accidental metaphysical part of it.[34]

To sum up: What explains a substance's having accidental *esse* is the metaphysical composition of that substance with an accidental form.[35] Before I get to Aquinas's explanation of a substance's having substantial *esse*, it is important to note that Aquinas acknowledges that a substance can have a multiplicity of accidental forms and consequently can have a multiplicity of accidental *esse*.[36] For example, let us assume there exists a dog which is called 'Fido'. Furthermore, let us assume that Fido exists as a barking, running, wet dog, weighing exactly 20 pounds. In Aquinas's view, what accounts for the existence of Fido as a barking, running, wet, and 20-pounds heavy dog is that the accidental forms of running, barking, wetness, and weighing 20 pounds inhere in Fido. They are accidental metaphysical parts of Fido by means of which he subsists as a dog which is barking, running, wet, and weighing 20 pounds.[37]

---

qualis vel quanta, non quod eis sit simpliciter sicut per formam substantialem: unde accidentia magis proprie dicuntur entis, quam entia, ut patet in Metaphys."

33 Cf. Brown, *Aquinas and the Ship of Theseus*, 54.

34 Whether accidental forms are metaphysical parts of substances is controversially discussed among contemporary Thomistic scholars. For a good overview of the recent debate and a defense of the claim that accidental forms are accidental metaphysical parts of substances, see Jeremy W. Skrzypek, "Accidental Forms as Metaphysical Parts of Material Substances in Aquinas's Ontology," *Oxford Studies in Medieval Philosophy* 7 (2019).

35 Since Aquinas also adheres to the thesis of the convertibility of being, i.e., *esse* in the sense of actuality, and goodness, this view has some interesting philosophical and theological implications because by the extension of their being through an accidental form substances also extent their goodness as has been explained in detail by Te Velde, *Participation and Substantiality in Thomas Aquinas*, 35–43; Gloria Frost, "Aquinas on the Intension and Remission of Accidental Forms," *Oxford Studies in Medieval Philosophy* 7 (2019).

36 Cf. *QQ*, IX, a. 2, corp.: "Et sic patet quod, secundum opinionem secundam, oportet dicere quod in Christo est unum esse substantiale, secundum quod esse proprie est suppositi, quamvis sit multiplex esse accidentale."

37 Cf. *QQ*, IX, a.2., corp.: "Omnibus vero quae non per se subsistunt sed in alio et cum alio, sive sint accidentia sive formae substantiales aut quaelibet partes, non habent esse ita quod ipsa vere sint, sed attribuitur eis esse alio modo, id est ut quo aliquid est, sicut albedo dicitur esse, non quia ipsa in esse subsistat, sed quia ea aliquid habet esse album."; *ST*, III, q. 17, a. 2, corp.: "Natura enim significatur per modum formae, quae dicitur ens ex eo quod ea aliquid est, sicut albedine est aliquid album, et humanitate est aliquis homo."; *QDUVI*, a. 4, corp.: "Esse enim proprie et vere dicitur de supposito subsistente. Accidentia enim et formae non subsistentes dicuntur esse, in quantum eis aliquid subsistit; sicut albedo dicitur ens, in quantum ea est aliquid album. Considerandum est autem, quod aliquae formae sunt quibus est aliquid ens non simpliciter, sed secundum quid; sicut sunt omnes formae accidentales. Aliquae autem formae

Let us now turn to Aquinas's explanation of a substance's having substantial *esse*. It is obvious that a substance is not actual as that which it is essentially by mediation of a form which is accidental to it. Fido does not subsist as a dog by mediation of a form such as barking, running, wetness, or weighing 20 pounds, which is accidental to him. For example, if Fido loses the accidental form of wetness, he ceases to exist as a wet dog, but he does not cease to exist as a dog. Furthermore, accidental forms can only inhere in Fido and be accidental metaphysical parts of Fido if Fido exists. Thus, only a substantial form can account for a substance's having substantial *esse*. In Aquinas's view, for a substance to exist is to have substantial *esse* by means of its substantial form or essence.[38] In the case of Fido, Fido has substantial *esse* by mediation of his substantial form or essence which is a dog's soul or dogness.

Thus, a first difference between accidental and substantial forms is precisely this: In virtue of having a substantial form, a substance has substantial *esse* and in virtue of having an accidental form a substance has accidental *esse*.[39] Consequently, a substantial form is not just an accidental metaphysical part of a substance like any accidental form of it but an essential metaphysical part of it. A substance can lose a particular accidental form it has and continue to exist, but it cannot lose the substantial form it has and continue to exist.

A second difference between accidental and substantial forms is that a substance can have a multiplicity of accidental forms but only one substantial form.[40] In Aquinas's view, the substantial form of a substance determines

sunt quibus res subsistens simpliciter habet esse; quia videlicet constituunt esse substantiale rei subsistentis."

38 Cf. *ST*, I, q. 76, a. 4, corp.: "Ad cuius evidentiam, considerandum est quod forma substantialis in hoc a forma accidentali differt, quia forma accidentalis non dat esse simpliciter, sed esse tale, sicut calor facit suum subiectum non simpliciter esse, sed esse calidum. Et ideo cum advenit forma accidentalis, non dicitur aliquid fieri vel generari simpliciter, sed fieri tale aut aliquo modo se habens, et similiter cum recedit forma accidentalis, non dicitur aliquid corrumpi simpliciter, sed secundum quid. Forma autem substantialis dat esse simpliciter, et ideo per eius adventum dicitur aliquid simpliciter generari, et per eius recessum simpliciter corrumpi." I will explain below the difference between a substantial form and the essence of a material substance. For the moment, we can ignore this difference.

39 Cf. *ST*, I, q. 50, a. 5, corp.: "Unumquodque enim est ens actu secundum quod habet formam; *DPN*, cap. 1: Sed duplex est esse, scilicet esse essentiale rei sive substantiale, ut hominem esse, et hoc est esse simpliciter; est autem aliud esse accidentale, ut hominem esse album, et hoc est esse aliquid. [...] Et quia forma facit esse in actu, ideo forma dicitur esse actus; quod autem facit actu esse substantiale est forma substantialis, et quod facit actu esse accidentale dicitur forma accidentalis."

40 Cf. *ST*, I, q. 76, a. 4, corp.: "Si igitur ita esset, quod praeter animam intellectivam praeexisteret quaecumque alia forma substantialis in materia, per quam subiectum animae esset ens

what *kind* of substance it is.[41] In virtue of having a substantial form, the substance is formally configured as that which it is and has the properties and causal powers which are characteristic for a member of a particular species.[42]

Before I can continue with my reconstruction and explanation of Aquinas's view that the existence of a substance consists in that substance's having *esse* by means of its essence or form, I shall say something about the difference between the essence of a substance and its substantial form. To explain the difference, it is helpful to have in mind that Aquinas distinguishes between three kinds of substantial forms. They can be distinguished according to the ways each of them has *esse*.

First, there are substantial forms which have *esse per se* and *in se*. Substantial forms which have *esse* in this way are substantial forms of immaterial substances like angels, i.e., substances whose substantial forms do not inform or configure prime matter.[43] These substantial forms are immaterial *substances* be-

---

actu; sequeretur quod anima non daret esse simpliciter; et per consequens quod non esset forma substantialis; et quod per adventum animae non esset generatio simpliciter, neque per eius abscessum corruptio simpliciter, sed solum secundum quid. Quae sunt manifeste falsa. Unde dicendum est quod nulla alia forma substantialis est in homine, nisi sola anima intellectiva; et quod ipsa, sicut virtute continet animam sensitivam et nutritivam, ita virtute continet omnes inferiores formas, et facit ipsa sola quidquid imperfectiores formae in aliis faciunt. Et similiter est dicendum de Anima sensitiva in brutis, et de nutritiva in plantis, et universaliter de omnibus formis perfectioribus respectu imperfectiorum."

41 Cf. *SCG*, IV, cap. 36: "Cuiuslibet enim naturae est aliqua operatio propria: nam forma est operationis principium, secundum quam unaquaeque natura habet propriam speciem."

42 Cf. *In Meta*, VII, lect. 11, 1519: "Sed animal videtur esse quoddam sensibile. Non enim potest definiri sine motu. Animal enim discernitur a non animali sensu et motu, ut patet in primo libro *De anima*. Et ideo non potest definiri animal sine partibus corporalibus habentibus se aliquo modo debito ad motum. Non enim manus est pars hominis quocumque modo se habens, sed quando est sic disposita quod potest perficere opus manus; quod non potest facere sine anima, quae est principium motus. Quare oportet quod manus cuiuscumque sit pars hominis, secundum quod est animata. Secundum vero quod est inanimata, non est pars, sicut manus mortua vel depicta. Unde oportet quod partes tales quae sunt necessariae ad perficiendum operationem speciei propriam, sint partes speciei; tam quae sunt ex parte formae, quam quae sunt ex parte materiae." For a more detailed explanation, see Stump, *Aquinas*, 36–38, 194–216.

43 Cf. *DEE*, cap. 4: "Unde oportet quod in qualibet substantia intelligente sit omnino immunitas a materia, ita quod neque habeat materiam partem sui, neque etiam sit sicut forma impressa in materia ut est de formis materialibus. [...] Unde illae formae quae sunt propinquissimae primo principio sunt formae per se sine materia subsistentes, non enim forma secundum totum genus suum materia indiget, ut dictum est."

cause they subsist with a complete essence or nature.[44] Consequently, each subsistent form which exists *per se* and *in se* (e.g., an angel) is its own species.[45]

Second, there are substantial forms which have *esse per se* but not *in se*. Substantial forms which have *esse* in this way are the substantial forms of human beings, i.e., human souls. A human soul which informs or configures prime matter subsists but its having *esse* is not dependent on matter.[46] In other words, a human soul can exist in separation from the matter it informs or configures.[47] Nevertheless, a human soul is not a substance because it does not subsist *in se*, i.e., it is not a subsistent thing with a complete essence or nature.[48] A human soul has a quiddity or essence which includes a reference to

44 Cf. *QQ*, II, q. 2, a. 2, ad. 1: "Substanciis vero inmaterialibus creatis accidunt quidem aliqua praeter rationem speciei quae non sunt determinativa essentialium principiorum, ut dictum est; non tamen accidunt eis aliqua quae sunt determinativa essentiae speciei, quia ipsa natura speciei non individuatur per materiam, sed per se ipsam, ex hoc quod talis forma non est nata recipi in aliqua materia; unde per se ipsam est non multiplicabilis neque praedicabilis de pluribus."
45 Cf. *ST*, I, q. 50, a. 4, corp.: "Si ergo angeli non sunt compositi ex materia et forma, ut dictum est supra, sequitur quod impossibile sit esse duos angelos unius speciei. Sicut etiam impossibile esset dicere quod essent plures albedines separatae, aut plures humanitates; cum albedines non sint plures nisi secundum quod sunt in pluribus substantiis."
46 Cf. *SCG*, II, cap. 81: "Ostensum est autem quod anima humana est forma secundum suum esse a materia non dependens."
47 Cf. *ST*, I, q. 75, a. 6, corp.: "Respondeo dicendum quod necesse est dicere animam humanam, quam dicimus intellectivum principium, esse incorruptibilem. Dupliciter enim aliquid corrumpitur, uno modo, per se; alio modo, per accidens. Impossibile est autem aliquid subsistens generari aut corrumpi per accidens, idest aliquo generato vel corrupto. Sic enim competit alicui generari et corrumpi, sicut et esse, quod per generationem acquiritur et per corruptionem amittitur. Unde quod per se habet esse, non potest generari vel corrumpi nisi per se, quae vero non subsistunt, ut accidentia et formae materiales, dicuntur fieri et corrumpi per generationem et corruptionem compositorum. Ostensum est autem supra quod animae brutorum non sunt per se subsistentes, sed sola anima humana. Unde animae brutorum corrumpuntur, corruptis corporibus, anima autem humana non posset corrumpi, nisi per se corrumperetur. Quod quidem omnino est impossibile non solum de ipsa, sed de quolibet subsistente quod est forma tantum. Manifestum est enim quod id quod secundum se convenit alicui, est inseparabile ab ipso. Esse autem per se convenit formae, quae est actus. Unde materia secundum hoc acquirit esse in actu, quod acquirit formam, secundum hoc autem accidit in ea corruptio, quod separatur forma ab ea. Impossibile est autem quod forma separetur a seipsa. Unde impossibile est quod forma subsistens desinat esse."
48 Cf. *ST*, I, q. 75, a.2, ad. 1: "Ad primum ergo dicendum quod hoc aliquid potest accipi dupliciter, uno modo, pro quocumque subsistente, alio modo, pro subsistente completo in natura alicuius speciei. Primo modo, excludit inhaerentiam accidentis et formae materialis, secundo modo, excludit etiam imperfectionem partis. Unde manus posset dici hoc aliquid primo modo, sed non secundo modo. Sic igitur, cum anima humana sit pars speciei humanae, potest

something extrinsic to it, namely, the human substance of which it is an essential metaphysical part.[49]

Finally, there are substantial forms which have neither *esse per se* nor *in se* but have *esse* only in another as a subject—they have *esse* only in matter—, i.e., their existence is dependent on matter. Substantial forms which have *esse* in this way are the substantial forms of all other material substances. These substantial forms inform or configure prime matter like the human soul but in contrast to the human soul they do not have *esse per se*, i.e., they do not subsist and cannot exist in separation from the matter they inform or configure.[50]

With the help of the distinction between these three kinds of substantial forms it is possible to understand why Aquinas distinguishes between the substantial form of a substance and its essence. What the substantial form and the essence of a substance have in common is that they are that by means of which (*quo est*) a substance has substantial *esse* and exists as what it is essentially.[51] In the case of an immaterial substance, there is no distinction between the essence of a substance and its substantial form.[52] Immaterial substances like angels are subsistent substantial forms, which means that they have a simple essence which does not contain matter.[53]

---

dici hoc aliquid primo modo, quasi subsistens, sed non secundo modo, sic enim compositum ex anima et corpore dicitur hoc aliquid."

49 Cf. Stump, *Aquinas*, 42.

50 Cf. *ST*, I, q. 75, a. 6, corp.: "Ostensum est autem supra quod animae brutorum non sunt per se subsistentes, sed sola anima humana. Unde animae brutorum corrumpuntur, corruptis corporibus, anima autem humana non posset corrumpi, nisi per se corrumperetur."

51 Cf. *In Meta*, V, lect. 10, 904: "Essentia enim et forma in hoc conveniunt quod secundum utrumque dicitur esse illud quo aliquid est. Sed forma refertur ad materiam, quam facit esse in actu; quidditas autem refertur ad suppositum, quod significatur ut habens talem essentiam. Unde sub uno comprehenduntur forma et species, idest sub essentia rei."; *DEE*, cap. 1: "Sed essentia dicitur secundum quod per eam et in ea ens habet esse."; *ST*, I, q. 50, a.5, corp.: "Unumquodque enim est ens actu secundum quod habet formam."; *SCG*, II, cap. 30: "Forma autem, secundum id quod est, actus est: et per eam res actu existunt."

52 Cf. *DEE*, cap. 4: "Unde illae formae quae sunt propinquissimae primo principio sunt formae per se sine materia subsistentes, non enim forma secundum totum genus suum materia indiget, ut dictum est; et huiusmodi formae sunt intelligentiae, et ideo non oportet ut essentiae vel quiditates harum substantiarum sint aliud quam ipsa forma." Nevertheless, there is a distinction between an immaterial substance and its essence or form—a supposit and its nature in Aquinas's terminology—due to the fact that a subsistent substantial form like an angel is not identical with its *esse*. Cf.: *QQ*, II, q. 2, a. 2, ad. 1: "Sed quia non est suum esse, accidit ei aliquid praeter rationem speciei, scilicet ipsum esse, et alia quaedam quae attribuuntur supposito et non naturae. Propter quod suppositum in eis non est omnino idem cum natura."; I will explain this point later in more detail when I get to the essence-*esse* composition of a substance.

53 Cf. *ST*, I, q. 50, a. 2; *QDSC*, a. 1.

In the case of a material substance, i.e., a substantial form which informs or configures prime matter, there is a difference between the substantial form and the essence of that substance, because a material substance is a *material* substance and since forms are immaterial, a material substance's substantial form alone cannot account for the *materiality* of a material substance. In Aquinas's view, material substances are essentially material, i.e., they are not substantial forms which are accidentally related to and united with matter.[54] Consequently, material substances must have an essence or form which contains a reference to matter to account for their materiality and the fact that they have a substantial form which informs or configures prime matter. Therefore, Aquinas distinguishes between the essence of a material substance as the form of the whole (*forma totius*) and its substantial form as the formal part of its essence (*forma partis*). For example, the essence or *forma totius* of a human being is humanity and not a human soul.[55] To sum up: Material substances have sub-

54 Cf. *In Meta*, VII, lect. 9, 1469: "Unde est alia opinio, quam sequitur Avicenna; et secundum hanc forma totius, quae est ipsa quidditas speciei, differt a forma partis, sicut totum a parte: nam quidditas speciei, est composita ex materia et forma, non tamen ex hac forma et ex hac materia individua. Ex his enim componitur individuum, ut Socrates et Callias. Et haec est sententia Aristotelis in hoc capitulo, quam introducit ad excludendum opinionem Platonis de ideis. Dicebat enim species rerum naturalium esse per se existentes sine materia sensibili, quasi materia sensibilis non esset aliquo modo pars speciei. Ostenso ergo, quod materia sensibilis sit pars speciei in rebus naturalibus, ostenditur quod impossibile est esse species rerum naturalium sine materia sensibili, sicut hominem sine carnibus et ossibus, et sic de aliis."; *ST*, I, q. 76, a. 1, a. 5, a. 6.

55 Cf. *In Meta*, VII, lect. 9, 1469: "Unde est alia opinio, quam sequitur Avicenna; et secundum hanc forma totius, quae est ipsa quidditas speciei, differt a forma partis, sicut totum a parte: nam quidditas speciei, est composita ex materia et forma, non tamen ex hac forma et ex hac materia individua. Ex his enim componitur individuum, ut Socrates et Callias."; *DEE*, cap. 2: "Nomen autem significans id unde sumitur natura speciei, cum praecisione materiae designatae, significat partem formalem. Et ideo humanitas significatur ut forma quaedam, et dicitur quod est forma totius; non quidem quasi superaddita partibus essentialibus, scilicet formae et materiae, sicut forma domus superadditur partibus integralibus eius: sed magis est forma quae est totum, scilicet formam complectens et materiam, tamen cum praecisione eorum per quae nata est materia designari." It is important to note that the matter which is included in the essence of a material substance is neither prime matter nor designated matter but non-designated matter, i.e., the type of matter out of which members of the species to which the relevant material substance belongs are made. I will get back to this point below. Cf. *DEE*, cap. 2: "Et ideo sciendum est quod materia non quolibet modo accepta est individuationis principium, sed solum materia signata; et dico materiam signatam quae sub determinatis dimensionibus consideratur. Haec autem materia in diffinitione quae est hominis in quantum est homo non ponitur, sed poneretur in diffinitione Socratis si Socrates diffinitionem haberet. In diffinitione autem hominis ponitur materia non signata: non enim in diffinitione hominis ponitur hoc os et haec caro, sed os et caro absolute quae sunt materia hominis non signata."

stantial *esse* by means of (*quo est*) a complex essence which contains a substantial form and matter.[56]

Thus, with the help of the exposition of Aquinas's views on *ens*, *esse*, form, and essence, I reconstructed a first preliminary answer to the nature question. According to Aquinas, the existence of a substance consists in a substance's having *esse*. Aquinas's explanation for a substance's 'having' *esse* is that a substance has *esse* by means of its essence or form. To be more precise, a substance has *substantial esse* by means of its essence or form.

6.2 The Existence of a Substance Consists in the Participation of Its Essence or Form in *Esse*

With the help of the exposition of Aquinas's views on *ens*, *esse*, form, and essence, I reconstructed a first preliminary answer to the nature question. According to Aquinas, the existence of a substance consists in a substance's having *esse*. And Aquinas's explanation for a substance's 'having' *esse* is that a substance has *esse* by means of its essence or form. To be more precise, a substance has *substantial esse* by means of its essence or form.

This is just the first stage of Aquinas's account of what it is for a substance to exist. What remains to be explained is how a substance has *esse* by means of its essence or form. There is something left to be explained because in Aquinas's view the *esse* of a substance cannot be identified with and reduced to a substance's essence or form.[57] In other words, a substance's having *esse* cannot be identified with and reduced to a substance's having an essence or form.

56 To be more precise, the matter which is part of the essence of a material substance is non-designated matter, i.e., the *type* of matter out of which the individual members of the species are made. I will explain this point below in more detail. Cf. *DEE*, cap. 4: "In hoc ergo differt essentia substantiae compositae et substantiae simplicis, quod essentia substantiae compositae non est tantum forma sed complectitur formam et materiam, essentia autem substantiae simplicis est forma tantum." In what follows, I will use the expression 'essence or form' for that by means of which a substance has *esse* to simplify the exposition as much as possible.

57 Cf. *QDV*, q. 21, a. 5: "Ipsa autem natura vel essentia divina est eius esse; natura autem vel essentia cuiuslibet rei creatae non est suum esse, sed est esse participans ab alio."; *DEE*, cap. 4: "Unde relinquitur quod talis res quae sit suum esse non potest esse nisi una; unde oportet quod in qualibet alia re praeter eam aliud sit esse suum et aliud quiditas vel natura seu forma sua; unde oportet quod in intelligentiis sit esse praeter formam, et ideo dictum est quod intelligentia est forma et esse."; *QQ*, II, q. 2, a. 2, corp.: "In solo autem Deo non invenitur aliquod accidens praeter eius essentiam, quia suum esse est sua essentia, ut dictum est, et ideo in Deo omnino est idem suppositum et natura. In angelo autem non est omnino idem, quia aliquid

Consequently, the existence of a substance does not simply consist in the substance's having an essence or form. According to Aquinas, a satisfactory explanation of what it is for a substance to exist has to take into account that a material substance requires a substantial form, matter, and *esse* for its existence, while an immaterial substance requires only a substantial form and *esse* for its existence.[58] Thus, saying that a substance has *esse* by means of its essence or form does not suffice to explain what it is for a substance to exist. If the essence or form of a substance is not identical with its *esse*, it must be explained how a substance, its essence or form, and its *esse* are related such that a substance has *esse* by means of its essence or form.

In what follows, I will argue that Aquinas makes use of the idea of participation to explain how a substance is related to its essence or form and its *esse* in

accidit ei praeter id quod est de ratione suae speciei, quia et ipsum esse angeli est praeter eius essentiam seu naturam et alia quaedam ei accidunt, quae omnia pertinent ad suppositum, non autem ad naturam."; *SCG*, II, cap. 54: "Tertio, quia nec forma est ipsum esse, sed se habent secundum ordinem: comparatur enim forma ad ipsum esse sicut lux ad lucere, vel albedo ad album esse. Deinde quia ad ipsam etiam formam comparatur ipsum esse ut actus. [...] Unde in compositis ex materia et forma nec materia nec forma potest dici ipsum quod est, nec etiam ipsum esse. Forma tamen potest dici quo est, secundum quod est essendi principium; ipsa autem tota substantia est ipsum quod est; et ipsum esse est quo substantia denominatur ens. In substantiis autem intellectualibus, quae non sunt ex materia et forma compositae, ut ostensum est, sed in eis ipsa forma est substantia subsistens, forma est quod est, ipsum autem esse est actus et quo est."; *QQ*, II, q. 2, a. 1, corp.: "Secundum hoc ergo dicendum est quod ens praedicatur de solo Deo essentialiter, eo quod esse divinum est esse subsistens et absolutum; de qualibet autem creatura praedicatur per participationem: nulla enim creatura est suum esse, sed est habens esse. [...] Quandocumque autem aliquid praedicatur de altero per participationem, oportet ibi aliquid esse praeter id quod participatur, et ideo in qualibet creatura est aliud ipsa creatura quae habet esse, et ipsum esse eius. Et hoc est quod Boethius dicit in libro De hebdomadibus, quod in omni eo quod est citra Primum, aliud est esse et quod est."

58 Cf. *SCG*, I, cap. 27: "Forma corporis non est ipsum esse, sed essendi principium."; *QDA*, q. 6, corp.: "In substantiis enim ex materia et forma compositis tria invenimus, scilicet materiam, et formam, et tertium esse, cuius quidem principium est forma, nam materia ex hoc quod recipit formam participat esse. Sic igitur esse consequitur ipsam formam, nec tamen forma est suum esse, cum sit eius principium. Et licet materia non pertingat ad esse nisi per formam, forma tamen, in quantum est forma, non indiget materia ad suum esse, cum ipsam formam consequatur esse, sed indiget materia cum sit talis forma quae per se non subsistat. Nihil igitur prohibet esse aliquam formam a materia separatam quae habeat esse; et in huiusmodi forma ipsa essentia formae comparatur ad esse sicut potentia ad proprium actum. Et ita in formis per se subsistentibus invenitur et potentia et actus, in quantum ipsum esse est actus formae subsistentis, quae non est suum esse." To be clear, immaterial substances also possess various accidents by means of which they have a multiplicity of accidental *esse*. I am thankful to Jeremy Skrzypek for making me aware of the need to clarify this point.

such a way that it has *esse* by means of its essence or form.[59] The second stage of Aquinas's account for what it is for a substance to exist is that that a substance has *esse* by means of its essence or form in virtue of the participation of its essence or form in *esse*.

6.2.1 The Core Idea of Aquinas's View: A Substance Is an *Ens* by Participation

Central to Aquinas's view about what it is for a substance to exist is the idea that a substance is not an *ens* by its essence or form (*per essentiam*) but by participation (*per participationem*).[60] To illustrate the difference, Aquinas performs a

---

59 Despite a wide agreement that participation plays an important and central role in Aquinas's metaphysics, there is an ongoing controversy among Thomistic scholars about which different ways or modes of participation Aquinas recognizes in general and in particular about how and in what way(s) something participates in *esse*. A list of some of the most important contributions to the debate includes Cornelio Fabro, "The Intensive Hermeneutics of Thomistic Philosophy: The Notion of Participation," *The Review of Metaphysics* 27 (1974); *Partecipazione e causalità secondo S. Tommaso D'Aquino*, 2 ed. (Turin: Società Editrice Internazionale, 1960); Louis Bertrand Geiger, *La participation dans la philosophie de s. Thomas d'Aquin*, 2 ed. (Paris: Librairie Philosophique J. Vrin, 1953); Te Velde, *Participation and Substantiality in Thomas Aquinas*; Wippel, *The Metaphysical Thought of Thomas Aquinas*, 94–131; "Thomas Aquinas and Participation," *Studies in Philosophy and the History of Philosophy* 17 (1987); William N. Clarke, "The Meaning of Participation in St. Thomas," *Proceedings of the American Catholic Philosophical Association* 26 (1952); Andrew Davison, *Participation in God: A Study in Christian Doctrine and Metaphysics* (Cambridge: Cambridge University Press, 2019). For reasons of space, I cannot engage with all the different readings and interpretations of Aquinas's view on participation which have been presented over the years. Such a comparative study would be an extensive book on its own. Rather, I will present my own reading and interpretation of Aquinas's view on how a substance participates in *esse* and limit myself to marking in the footnotes where my understanding of Aquinas's views diverges from some important contemporary interpreters like Wippel or Te Velde. Maybe the most important contribution of my interpretation to the debate is that I reconstruct Aquinas's view on the participatory relationship between a substance, its essence or nature, and its *esse* in terms of a relation between a determinate, a determinable, and a determinant. To the best of my knowledge, no interpreter has ever tried before to make sense of Aquinas's view on a substance's participatory relationship to its *esse* in this way.

60 Cf. *SCG*, II, cap. 15: "Deus autem est ens per essentiam suam: quia est ipsum esse. Omne autem aliud ens est ens per participationem: quia ens quod sit suum esse non potest esse nisi unum ut in primo ostensum est."; *SCG*, III, cap. 66: "Deus autem solus est ens per essentiam suam, omnia autem alia sunt entia per participationem: nam in solo Deo esse est sua essential."; *QQ*, II, q. 2, a. 1, corp.: "Secundum hoc ergo dicendum est quod ens praedicatur de solo Deo essentialiter, eo quod esse divinum est esse subsistens et absolutum; de qualibet autem creatura praedicatur per participationem: nulla enim creatura est suum esse, sed est habens esse."

thought experiment: Let us suppose for the sake of explanation that there exists some separate light, i.e., that light as a form subsists in separation from matter. In this case, there exists an *ens* which is light by its essence or form.[61] Such a separate form does not *have* light, i.e., it is not illuminated through the form of light, which is added to it, but *is* light itself. In other words, the correct answer to the question 'What is it?' is that it is light because it is identical with the essence or form of light. Thus, if light subsists in separation from matter, it exists as an *ens* which is essentially light and therefore pure light.

Now, let us suppose there exists an illuminated body, for example, a dog called 'Fido' who is playing in the sun and compare Fido with light as a subsistent form. It would not only be awkward but also false to say that Fido is light. Fido is not essentially light but something which *has* light, i.e., he is something which is illuminated through the form of light which is added to him. The fact that Fido is an *ens* which is not light but has light, i.e., the fact that he is illuminated and not light itself, rules out that Fido is an *ens* which is identical with the essence or form of light. As that which he is, Fido is not identical with the essence or form of light but has a different essence or form. The correct answer to the question 'What is Fido?' is that Fido is a dog. Fido is a dog because his essence or form is dogness and not light. Consequently, Fido subsists as an *ens* whose essence is not identical with the essence or form of light. Nevertheless, Fido subsists not only as a dog but also as a dog which has light, i.e., as an illuminated dog. What accounts for this fact is that he is an *ens* which has light by participation, i.e., that he subsists as an *ens* to which the form of light is added which participates in the sunlight's power to illuminate.

It is important to note that Aquinas is not committed to the existence of subsistent forms such as light or whiteness. He only uses the hypothetical case of a subsistent form such as light or whiteness to illustrate the difference between an *ens* which is an *ens* by its essence and an *ens* which is an *ens* by participation. An *ens* which is an *ens* by its essence is an *ens* which does not have *esse* but is *esse* because its essence is *esse*. Such an *ens* subsists as an *ens* whose essence is identical with *esse*. In other words, such an *ens* is subsistent *esse* itself, i.e., it is pure *esse*.[62]

61 Cf. *QQ*, II, q. 2, a. 1, corp.: "Dicendum quod dupliciter aliquid e aliquo praedicatur: uno modo essentialiter, alio modo per participationem: lux enim praedicatur de corpore illuminato participative, sed, si esset aliqua lux separata, praedicaretur de ea essentialiter."

62 Cf. *QQ*, II, q. 2, a. 1, corp.: "Secundum hoc ergo dicendum est quod ens praedicatur de solo Deo essentialiter, eo quod esse divinum est esse subsistens et absolutum; de qualibet autem creatura praedicatur per participationem: nulla enim creatura est suum esse, sed est habens esse."; *DSS*, cap. 8: "Omne autem quod est esse habet; est igitur in quocumque praeter primum

In contrast, an *ens* by participation is an *ens* which is not *esse* but which has *esse* because its essence is not *esse*. Such an *ens* subsists as an *ens* with an essence that is not identical with *esse*. For example, it is false to say that Fido is *esse* or subsists as *esse*. Fido is not subsistent *esse* itself or pure *esse* but *esse* which is limited and contracted to the *esse* that members of the species *dog* have. The correct answer to the question 'What is Fido?' is that Fido is a dog. Fido is a dog because his essence or form is dogness and not *esse*. Consequently, Fido subsists as an *ens* whose essence is not identical with *esse*. Nevertheless, Fido subsists, i.e., he has *esse* as a dog. What accounts for this fact is that he is an *ens* which has *esse* by participation, i.e., that he subsists as an *ens* which participates in *esse*.[63]

To sum up: An *ens* which is *esse* in its entirety does not participate in *esse* but is an *ens* by its essence, i.e., it is an *ens* whose essence is *esse*, whereas an *ens* which is not *esse* in its entirety but only has *esse* in addition to or joined to what it is by its essence is an *ens* by participation, i.e., it is an *ens* whose essence is not identical with *esse*.[64]

6.2.2 How a Substance Participates in *Esse*

The upshot of the previous section was that no substance, i.e., no *ens* which belongs to the genus *substance*, is an *ens* by its essence. Every *ens* which belongs to the genus *substance* is an *ens* by participation. It is obvious that no *ens* which belongs to the genus *substance* is an *ens* by its essence because no substance in its entirety is *esse*. Consequently, every *ens* which belongs to the genus *substance* must be an *ens* by participation, i.e., an *ens* which has *esse* in virtue of participating in *esse*. So, what remains to be explained is *how* a substance participates in *esse*. In particular, what needs to be explained is how an *ens* has *esse*

---

et ipsum esse tanquam actus, et substantia rei habens esse tanquam potentia receptiva huius actus quod est esse."; *SCG*, I, cap. 22: "Omnis res est per hoc quod habet esse."; ibid.: "Nulla igitur res cuius essentia non est suum esse, est per essentiam suam, sed participatione alicuius, scilicet ipsius esse."; *QDV*, q. 21, a. 5, corp.: "Et sic in Deo est esse purum, quia ipse Deus est suum esse subsistens; in creatura autem est esse receptum vel participatum."

63 Cf. *QQ*, II, q. 2, a. 1, corp.: "Quandocumque autem aliquid praedicatur de altero per participationem, oportet ibi aliquid esse praeter id quod participatur, et ideo in qualibet creatura est aliud ipsa creatura quae habet esse, et ipsum esse eius. Et hoc est quod Boethius dicit in libro De hebdomadibus, quod in omni eo quod est citra Primum, aliud est esse et quod est."

64 Cf. *In Meta*, I, lect. 10, 154: "Quod enim totaliter est aliquid, non participat illud, sed est per essentiam idem illi. Quod vero non totaliter est aliquid habens aliquid aliud adiunctum, proprie participare dicitur."

in addition to or adjoined to what it is by its essence without it being the case that *esse* is accidental to it.[65]

Before I proceed, it is important to have in mind that Aquinas distinguishes between two different kinds of participation in *esse*. According to each mode of participation there is something which receives in a particular way what belongs to that in which it participates in a universal way.[66] The two modes of participation differ with respect to whether the *esse* which is participated in is extrinsic or intrinsic to the participating substance. A substance participates in a different way in *esse* which is extrinsic to it than in *esse* which is intrinsic to it.

In the former case, the *esse* in which a substance participates is subsistent *esse* itself and a substance participates in it like an effect participates in the power of its cause.[67] In other words, subsistent *esse* itself as extrinsic

---

65 I am thankful to John Peck for making me aware of the need to improve the formulation of a previous version of this passage.

66 Cf. *In BDH*, lect. 2, 24: "Est autem participare quasi partem capere. Et ideo quando aliquid particulariter recipit id quod ad alterum pertinet universaliter, dicitur participare illud, sicut homo dicitur participare animal quia non habet rationem animalis secundum totam communitatem; et eadem ratione Socrates participat hominem."; *DSS*, cap. 3: "Omne autem participans aliquid accipit id quod participat ab eo a quo participat, et quantum ad hoc id a quo participat est causa ipsius: sicut aer habet lumen participatum a sole, quae est causa illuminationis ipsius."; *In CA*, lect. 4, 109: "Si autem aliquid sic haberet infinitam virtutem essendi quod non participaret esse ab alio, tunc esset solum infinitum; et tale est Deus, ut dicitur infra in XVI propositione. Sed, si sit aliquid quod habeat infinitam virtutem ad essendum secundum esse participatum ab alio, secundum hoc quod esse participat est finitum, quia quod participatur non recipitur in participante secundum totam suam infinitatem sed particulariter."; ibid., 146: "Ipsum autem esse participatum vocat finitum quia non participatur secundum totam infinitatem suae universalitatis sed secundum modum naturae participantis."; *SCG*, I, cap. 32: "De Deo autem nihil dicitur per participationem: nam omne quod participatur determinatur ad modum participati, et sic partialiter habetur et non secundum omnem perfectionis modum."

67 Cf. *In BDH*, lect. 2, 24: "Et similiter etiam effectus dicitur participare suam causam, et praecipue quando non adaequat virtutem suae causae, puta si dicamus quod aer participat lucem solis quia non recipit eam in claritate qua est in sole."; *DSS*, cap. 3: "Omne autem participans aliquid accipit id quod participat ab eo a quo participat, et quantum ad hoc id a quo participat est causa ipsius: sicut aer habet lumen participatum a sole, quae est causa illuminationis ipsius."; *In CA*, lect. 4, 109: "Si autem aliquid sic haberet infinitam virtutem essendi quod non participaret esse ab alio, tunc esset solum infinitum; et tale est Deus, ut dicitur infra in XVI propositione. Sed, si sit aliquid quod habeat infinitam virtutem ad essendum secundum esse participatum ab alio, secundum hoc quod esse participat est finitum, quia quod participatur non recipitur in participante secundum totam suam infinitatem sed particulariter."; ibid., 146: "Ipsum autem esse participatum vocat finitum quia non participatur secundum totam infinitatem suae universalitatis sed secundum modum naturae participantis."; *SCG*, I, cap. 32: "De Deo autem nihil dicitur per participationem: nam omne quod participatur determinatur ad modum participati, et sic partialiter habetur et non secundum omnem perfectionis modum."; *In DDN*,

and uncreated *esse* is the efficient and exemplary cause of the created *esse* which is intrinsic to a substance.[68] The *esse* which a substance has and which is intrinsic to it bears a certain likeness or similitude to subsistent *esse* itself which is extrinsic to it precisely because the latter is the efficient and exemplary cause of the former.[69] Nevertheless, the intrinsic *esse* of any substance is fundamentally distinct from subsistent *esse* itself or divine *esse*, which is extrinsic to it because the former is *esse* which does not exclude the possibility that something is added to it, while the latter *esse* is *esse* which excludes the possibility that something is added to it.[70] Strictly speaking, a

---

cap. 5, lect. 1, 629: "Omnia autem alia, sicut superius dictum est, habent esse receptum et participatum et ideo non habent esse secundum totam virtutem essendi, sed solus Deus, qui est ipsum esse subsistens, secundum totam virtutem essendi, esse habet."

68 Cf. *QDV*, q. 21, a. 4, corp.: "Hoc enim Deum omnes de Deo loquentes intelligunt, quod est omnium principium effectivum, cum oporteat omnia entia ab uno primo ente effluere. [...] Unde si prima bonitas sit effectiva omnium bonorum, oportet quod similitudinem suam imprimat in rebus effectis; et sic unumquodque dicetur bonum sicut forma inhaerente per similitudinem summi boni sibi inditam, et ulterius per bonitatem primam, sicut per exemplar et effectivum omnis bonitatis creatae. Et quantum ad hoc opinio Platonis sustineri potest. Sic igitur dicimus secundum communem opinionem, quod omnia sunt bona creata bonitate formaliter sicut forma inhaerente, bonitate vero increata sicut forma exemplari."; *In DDN*, cap. 5, lect. 1, 630–631: "Et quod plus est, neque est, secundum scilicet quod significatur tempus praesens, quia eius esse tempore non mensuratur, sed ipse est esse existentibus, non quidem ita quod ipse Deus sit esse formale existentium, sed eo modo loquendi utitur quo Platonici utebantur qui esse separatum dicebant esse existentium, inquantum compositiva per participationem abstractorum participantur. [...] Deinde, cum dicit: resumentes etc., colligit ea quae dicta sunt et dicit resumendo quod omnia existentia et mensurae essendi habent esse a praeexistente; et omne aevum et tempus, quae sunt durationes, sunt ex ipso et ipse est principiumeffectivum et causa finalis omnis saeculi et temporis et cuiuslibet quocumque modo existentis; et iterum omnia ipso participant, sicut prima forma exemplari; et non solum est causa quantum ad fieri rerum, sed et quantum ad totum esse et durationem, quod manifestat cum dicit: et a nullo existentium recedit: aedificatore enim recedente, domus remanet, quia est causa domus quantum ad fieri et non quantum ad esse, sed si Deus ab effectu recederet, effectus non remaneret, quia est causa ipsius esse. Et quia causa praeeminet effectibus, ipse est ante omnia et omnia consistunt in ipso, Colos. I, sicut effectus virtute praeexistunt in causa; et universaliter, quidquid quocumque modo est, praeexistit in primo ente, scilicet Deo et quantum ad esse quod habet in intellectu et quantum ad conservationem sui esse."

69 For a more detailed explanation of this point, see Gregory T. Doolan, "Aquinas on *Esse Subsistens* and the Third Mode of Participation," *The Thomist* 82, no. 4 (2018); *Aquinas on the Divine Ideas as Exemplar Causes* (Washington, D.C.: The Catholic University of America Press, 2008), 156–243.

70 Cf. *QDV*, q. 21, a. 4, ad. 7: "Ad septimum dicendum, quod cum dicitur: esse est proprium Deo; non est intelligendum quod nullum aliud esse sit nisi increatum; sed quod solum illud esse proprie dicitur esse, in quantum ratione suae immutabilitatis non novit fuisse vel futurum esse.

substance participates in divine *esse* only by a certain likeness or similitude which its intrinsic *esse* bears to subsistent *esse* itself as its cause.[71]

Aquinas's view on how a substance participates in divine *esse* or subsistent *esse* itself provides an answer to the strong ground question, i. e., it answers the question why there is something rather than nothing. The participation of a substance in divine *esse* explains why a substance has *esse* and exists although it might as well not have *esse* and not exist. Any *ens* which is an *ens* by participation has *esse* and exists because it stands in a causal relation to subsistent *esse* itself which is extrinsic to it and in virtue of which it has the *esse* it has which is intrinsic to it.[72]

In this inquiry, I am not concerned with the strong ground question. I am not trying to answer the question of why a substance has the *esse* it has. For this rea-

---

Esse autem creaturae dicitur esse per quamdam similitudinem ad illud primum esse, cum habeat permixtionem eius quod est futurum esse vel fuisse, ratione mutabilitatis creaturae. Vel potest dici, quod esse est proprium Deo, quia solus Deus est suum esse; quamvis alia esse habeant, quod esse non est esse divinum."; *QDP*, VII, a. 2, ad. 4: "Ad quartum dicendum, quod esse divinum, quod est eius substantia, non est esse commune, sed est esse distinctum a quolibet alio esse. Unde per ipsum suum esse Deus differt a quolibet alio ente."; ibid. ad. 6: "Ad sextum dicendum, quod ens commune est cui non fit additio, de cuius tamen ratione non est ut ei additio fieri non possit; sed esse divinum est esse cui non fit additio, et de eius ratione est ut ei additio fieri non possit; unde divinum esse non est esse commune."

71 Cf. *QQ*, XII, q. 4, a. 1, corp.: "Sciendum ergo quod unumquodque quod est in potentia et in actu, fit actu per hoc quod participat actum superiorem; per hoc autem aliquid maxime fit actu, quod participat per similitudinem primum et purum actum; primus autem actus est esse subsistens per se; unde completionem unumquodque recipit per hoc quod participat esse. Unde esse est completivum omnis formae, quia per hoc completur quod habet esse, et habet esse cum est actu; et sic nulla forma est nisi per esse."

72 Cf. *ST*, I, q. 44, a. 1, corp.: "Respondeo dicendum quod necesse est dicere omne quod quocumque modo est, a Deo esse. Si enim aliquid invenitur in aliquo per participationem, necesse est quod causetur in ipso ab eo cui essentialiter convenit; sicut ferrum fit ignitum ab igne. Ostensum est autem supra, cum de divina simplicitate ageretur, quod Deus est ipsum esse per se subsistens. Et iterum ostensum est quod esse subsistens non potest esse nisi unum, sicut si albedo esset subsistens, non posset esse nisi una, cum albedines multiplicentur secundum recipientia. Relinquitur ergo quod omnia alia a Deo non sint suum esse, sed participant esse. Necesse est igitur omnia quae diversificantur secundum diversam participationem essendi, ut sint perfectius vel minus perfecte, causari ab uno primo ente, quod perfectissime est."; ibid., ad. 1: Ad primum ergo dicendum quod, licet habitudo ad causam non intret definitionem entis quod est causatum, tamen sequitur ad ea qua sunt de eius ratione, quia ex hoc quod aliquid per participationem est ens, sequitur quod sit causatum ab alio. Unde huiusmodi ens non potest esse, quin sit causatum; sicut nec homo, quin sit risibile. Sed quia esse causatum non est de ratione entis simpliciter, propter hoc invenitur aliquod ens non causatum."; *SCG*, II, cap. 15: "Quod per essentiam dicitur, est causa omnium quae per participationem dicuntur: sicut ignis est causa omnium ignitorum inquantum huiusmodi."

son, I do not pursue any further the question of how Aquinas explains the participation of a substance in the *esse* which is extrinsic to it.[73] Rather, given that something exists, and that Aquinas's view is that that which exists exists in virtue of having *esse*, I am interested in the question of how this 'having' works according to Aquinas. This question is not satisfactorily answered by Aquinas's view that a substance participates in *esse* which is *extrinsic* to it, because it is still left unexplained how a substance 'has' the caused *esse* which is *intrinsic* to it. Nor is it satisfactorily answered by Aquinas's view that a substance has the *esse* which is intrinsic to it in virtue of participating in it, unless it is explained in detail how a substance participates in this *esse* which is intrinsic to it.[74]

Aquinas begins the exposition of how a substance participates in the *esse* which is intrinsic to it with the observation that whenever something is partici-

---

73 A helpful exposition is provided, for example, by Wippel, *The Metaphysical Thought of Thomas Aquinas*, 94–131.

74 It is important not to confuse the strong ground question with the nature question. The distinction is subtle but there is a difference. The nature question asks what it is for an object to have existence, i.e., the question is interested in the way in which 'having existence' works. Instead, the strong ground question asks about the source or cause of the existence that existing objects have. Why do existing objects have what they do not have simply in virtue of what they are? Maybe, the following analogy will be helpful to illustrate the difference. It is one thing to ask for a metaphysical explanation for a dove's having the property of wearing a ring around its foot. A possible answer to this question is that the dove wearing the ring is metaphysically composed of a dove and the property of wearing a ring. It is another thing to ask for a metaphysical explanation for the cause of the dove's having the property of wearing a ring around its foot. A possible answer to this question is that a human being caused the dove to have the property (form) of wearing a ring. Now, Aquinas's answer to the strong ground question is that any *ens* which is an *ens* by participation has *esse* and exists because it stands in a causal relation to subsistent *esse* itself which is *extrinsic* to it, in virtue of which it has the *esse* which is *intrinsic* to it. In short, an *ens* by participation has the *esse* it has because it participates in divine *esse*. But since I am not attempting to answer the strong ground question in this essay, I do not need to explain how this kind of participation is supposed to work. All I need to explain is how a substance participates in the *esse* which is *intrinsic* to it because this is the only kind of participation which plays a role in Aquinas's answer to the nature question. And this is what I am going to explain in the following. Thus, everything that needs to be explained gets explained. The same does not hold for the property view, the unrestricted-domain view, and the restricted-domain view because their answers to the *nature question* involve a reference to something extrinsic to them. So, it is not unfair to object that the explanations these views offer are unsatisfactory unless they explain what it is for that which is extrinsic to them to exist because a reference to an extrinsic entity plays a role in their answers to the nature question. However, such a reference to an extrinsic entity plays no role in Aquinas's answer to the nature question. It plays a role only in Aquinas's answer to the strong ground question. I am thankful to Timothy Pawl for raising this worry.

pated in there must be something in addition to or besides that which is participated in, namely, that which participates.[75] For example, if there is an illuminated body, i.e., a body which participates in light, there must be something in addition to light which is participated in, namely, the body which participates in the light. Thus, in any case of participation there must be something which participates (*quod est*) and something else which is participated in and in virtue of which (*quo est*) that which participates has what it does not have in virtue of itself.

Thus, in the case of an *ens* by participation, i.e., a substance which participates in the *esse* which is intrinsic to it, there must be something which participates and something else which is participated in, i.e., there must be that which has and participates (*quod est*) in *esse* and the *esse* which is participated in (*quo est*).[76] In other words, any substance is a participant (*participans*) in *esse* which is metaphysically composed of that which participates in *esse* (*participante*) and the *esse* it participates in (*participato*), which are related as potentiality to actuality.[77]

Thus, a substance participates in a different way in the *esse* which is intrinsic to it than in the *esse* which is extrinsic to it. The participation is a participation by composition only in the former case. In other words, only in the former case is that which participates added to the *esse* which is participated in in such a way that yields a metaphysical composite of that which participates (*quod est*) and *esse*.

So far, we have seen that the participation of a substance in the *esse* which is intrinsic to it is a participation by composition in contrast to its participation in

---

75 Cf. *QQ*, II, q. 2, a. 1, corp.: "Quandocumque autem aliquid praedicatur de altero per participationem, oportet ibi aliquid esse praeter id quod participatur."

76 Cf. *QQ*, II, q. 2, a. 1, corp.: "Quandocumque autem aliquid praedicatur de altero per participationem, oportet ibi aliquid esse praeter id quod participatur, et ideo in qualibet creatura est aliud ipsa creatura quae habet esse, et ipsum esse eius. Et hoc est quod Boethius dicit in libro De hebdomadibus, quod in omni eo quod est citra Primum, aliud est esse et quod est." Cf. *SCG*, II, cap. 54: "In substantiis autem intellectualibus, quae non sunt ex materia et forma compositae, ut ostensum est, sed in eis ipsa forma est substantia subsistens, forma est quod est, ipsum autem esse est actus et quo est. Et propter hoc in eis est unica tantum compositio actus et potentiae, quae scilicet est ex substantia et esse, quae a quibusdam dicitur ex quod est et esse; vel ex quod est et quo est. In substantiis autem compositis ex materia et forma est duplex compositio actus et potentiae: prima quidem ipsius substantiae, quae componitur ex materia et forma; secunda vero ex ipsa substantia iam composita et esse, quae etiam potest dici ex quod est et esse; vel ex quod est et quo est."

77 Cf. *In Phy*, VIII, lect. 21, 1153: "Omnis ergo substantia quae est post primam substantiam simplicem, participat esse. Omne autem participans componitur ex participante et participato, et participans est in potentia ad participatum."

the *esse* which is extrinsic to it, which can be called a participation by likeness or similitude.[78] In order to explain how a substance participates in the *esse* which is intrinsic to it, Aquinas distinguishes between two ways in which something is participated in. Something is participated in as something which is included in the essence or nature of that which participates, or something is participated in as something which is not included in the essence or nature of that which participates.[79]

Let us begin with the former case. In this way, a genus like *animal* is participated in by a species like *rational animal* and a species like *rational animal* is participated in by an individual of that species such as Joe Biden. An individual human being like Joe Biden participates in the species *rational animal* and the species *rational animal* participates in the genus *animal* because the individual does not possess the intelligible content of the species *rational animal* according to its full universality (*secundum totam communitatem*) and the species *rational animal* does not possess the intelligible content of the genus *animal* according to its full universality.[80] The species *rational animal* does not possess the intelligi-

78 The terms 'participation by composition' and 'participation by likeness or similitude' have been introduced by Fabro and Geiger in order to reconstruct Aquinas's view on participation, cf. Geiger, *La participation dans la philosophie de s. Thomas d'Aquin*; Fabro, *Partecipazione e causalità secondo S. Tommaso D'Aquino*. There is an ongoing controversy about what role these two different kinds or modes of participation play in Aquinas's view on how a substance participates in *esse*. I am following Wippel in his attempt to reconcile Fabro's emphasis on participation by composition with Geiger's emphasis on participation by likeness or similitude. Where I part company with Wippel and agree with Te Velde is that a substance's participation in the *esse* which is intrinsic to it is not a case of participation where an effect participates in the power of its cause, cf. Te Velde, *Participation and Substantiality in Thomas Aquinas*, 79; Wippel, *The Metaphysical Thought of Thomas Aquinas*, 94–131, see especially 109.
79 Cf. *QQ*, II, q. 2, a. 1, corp.: "Sed sciendum est quod aliquid participatur dupliciter. Uno modo quasi existens de substantia participantis, sicut genus participatur a specie; hoc autem modo esse non participatur a creatura: id enim est de substantia rei quod cadit in eius definitione, ens autem non ponitur in definitione creaturae, quia nec est genus neque est differentia. Unde participatur sicut aliquid non existens de essentia rei, et ideo alia quaestio an est et quid est; unde, cum omne quod est praeter essentiam rei dicatur accidens, esse, quod pertinet ad quaestionem an est, est accidens."
80 Cf. *In BDH*, lect. 2, 24: "Et ideo quando aliquid particulariter recipit id quod ad alterum pertinet universaliter, dicitur participare illud, sicut homo dicitur participare animal quia non habet rationem animalis secundum totam communitatem; et eadem ratione Socrates participat hominem."; *QQ*, II, q. 2, a. 1, corp.: "Sed sciendum est quod aliquid participatur dupliciter. Uno modo quasi existens de substantia participantis, sicut genus participatur a specie; hoc autem modo esse non participatur a creatura: id enim est de substantia rei quod cadit in eius definitione, ens autem non ponitur in definitione creaturae, quia nec est genus neque est differentia."; *SCG*, I, cap. 32: "Omne quod de pluribus praedicatur univoce, secundum participationem cuili-

ble content of the genus *animal* according to its full universality because members of the species *rational animal* only realize a particular and limited way of being an animal. For example, the genus *animal* contains species of animals which can fly or breathe under water, i.e., it contains possible ways of being an animal which are not realized by members of the species *rational animal.*

In the same way, an individual of the species *rational animal* does not possess the intelligible content of the species *rational animal* according to its full universality. Different individual human beings only realize particular and limited ways of being a human being. Joe Biden is a human being or rational animal because he has humanity, but he is not humanity.[81] Furthermore, as I will explain in more detail below, any individual human being participates in its species because any such individual is not identical with the common nature as expressed by the species but has something else—designated matter, i.e., matter considered under determinate dimensions—which individuates it so that it differs from all other members of the same species.[82]

There are at least three reasons why Aquinas rejects the possibility that a substance participates in the *esse* which is intrinsic to it in the way a genus is participated in by a species or a species by an individual member of that species.[83] First, a substance does not participate in the *esse* which is intrinsic to it in this way because a genus is participated in by a species and a species is participated in by an individual member of that species only in a logical but not in a real or ontological sense.[84] In both cases, a more universal or extended intelligi-

---

bet eorum convenit de quo praedicatur: nam species participare dicitur genus, et individuum speciem." See also Wippel, *The Metaphysical Thought of Thomas Aquinas*, 97.

81 Cf. *SCG*, I, cap. 32: "De aliis autem praedicationes fiunt per participationem, sicut Socrates dicitur homo non quia sit ipsa humanitas, sed humanitatem habens."

82 Cf. *DEE*, cap. 2: "Et ideo sciendum est quod materia non quolibet modo accepta est individuationis principium, sed solum materia signata; et dico materiam signatam quae sub determinatis dimensionibus consideratur. Haec autem materia in diffinitione quae est hominis in quantum est homo non ponitur, sed poneretur in diffinitione Socratis si Socrates diffinitionem haberet. In diffinitione autem hominis ponitur materia non signata: non enim in diffinitione hominis ponitur hoc os et haec caro, sed os et caro absolute quae sunt materia hominis non signata." See also Te Velde, *Participation and Substantiality in Thomas Aquinas*, 11–12.

83 For a different interpretation of Aquinas's view on the first mode of participation and an attempt to rebut the arguments which are about to follow, see Jason A. Mitchell, "Aquinas on *Esse Commune* and the First Mode of Participation," *The Thomist* 82, no. 4 (2018). For a general analysis and discussion of the three modes of participation that Aquinas mentions in his commentary on Boethius's *De hebdomadibus*, see Stephen L. Brock, "Harmonizing Plato and Aristotle on *Esse:* Thomas Aquinas and the *De hebdomadibus*," *Nova et Vetera* 5, no. 3 (2007).

84 Cf. *QDP*, q. 7, a. 6, corp.: "Sunt autem quaedam rationes quibus in re intellecta nihil respondet; sed ea quorum sunt huiusmodi rationes, intellectus non attribuit rebus prout in se ipsis

ble content is participated in by a less extended intelligible content.[85] This participation of an individual in its species and a species in its genus is not a case of a real or ontological participation but only a case of logical participation, as Aquinas rejects the Platonic view that a species exists by itself separately from its members and that a genus exists by itself separately from its species.[86]

Second, since Aquinas rejects the Platonic idea that a genus and a specific difference exist by themselves as separated ideas or forms, there is no opposition between being something by participation and being something by essence. In the case of a logical participation, what is said of something by participation can also be predicated of it substantially. The fact that Joe Biden is not humanity but only has humanity does not conflict with the fact that Joe Biden is essentially a human being or rational animal.[87] Since the genus *animal* is participated in by the species *rational animal* as something with is included in the essence or na-

---

sunt, sed solum prout intellectae sunt; sicut patet in ratione generis et speciei, et aliarum intentionum intellectualium: nam nihil est in rebus quae sunt extra animam, cuius similitudo sit ratio generis vel speciei. Nec tamen intellectus est falsus: quia ea quorum sunt istae rationes, scilicet genus et species, non attribuit rebus secundum quod sunt extra animam, sed solum secundum quod sunt in intellectu. Ex hoc enim quod intellectus in se ipsum reflectitur, sicut intelligit res existentes extra animam, ita intelligit eas esse intellectas: et sic, sicut est quaedam conceptio intellectus vel ratio, —cui respondet res ipsa quae est extra animam— ita est quaedam conceptio vel ratio, cui respondet res intellecta secundum quod huiusmodi; sicut rationi hominis vel conceptioni hominis respondet res extra animam; rationi vero vel conceptioni generis aut speciei, respondet solum res intellecta."

85 Cf. Wippel, *The Metaphysical Thought of Thomas Aquinas*, 97.

86 Cf. DEE, cap. 3: "Quia autem id cui convenit ratio generis vel speciei vel differentiae praedicatur de hoc singulari signato, impossibile est quod ratio universalis, scilicet generis vel speciei, conveniat essentiae secundum quod per modum partis significatur, ut nomine humanitatis vel animalitatis; et ideo dicit Avicenna quod rationalitas non est differentia sed differentiae principium; et eadem ratione humanitas non est species, nec animalitas genus. Similiter etiam non potest dici quod ratio generis vel speciei conveniat essentiae secundum quod est quaedam res exsistens extra singularia, ut Platonici ponebant, quia sic genus et species non praedicarentur de hoc individuo; non enim potest dici quod Socrates sit hoc quod ab eo separatum est, nec iterum illud separatum proficeret in cognitionem huius singularis. Et ideo relinquitur quod ratio generis vel speciei conveniat essentiae secundum quod significatur per modum totius, ut nomine hominis vel animalis, prout implicite et indistincte continet totum hoc quod in individuo est." See also Te Velde, *Participation and Substantiality in Thomas Aquinas*, 12.

87 Cf. *In BDH*, lect. 3, 45: "Sed in alio participationis modo, quo scilicet species participat genus, hoc etiam verum est secundum sententiam Platonis qui posuit aliam esse ideam animalis et bipedis et hominis; sed secundum Aristotelis sententiam qui posuit quod homo vere est id quod est animal, quasi essentia animalis non existente praeter differentiam hominis, nihil prohibet id quod per participationem dicitur etiam substantialiter praedicari."; See also Te Velde, *Participation and Substantiality in Thomas Aquinas*, 13 and Wippel, *The Metaphysical Thought of Thomas Aquinas*, 105.

ture of the species *rational animal*—the genus *animal* is part of the essence or nature of the species *rational animal*—, a species like *rational animal* is essentially what the genus *animal* is in which it participates.[88] In other words, the essence of Joe Biden, the essence of the species *rational animal*, and the essence of the genus *animal* only differ as the designated or determined differs from the undesignated or undetermined. The species *rational animal* determines the genus *animal* through the constitutive difference *rational*, and an individual human being determines the species *rational animal* through matter determined by dimensions.[89]

Finally, Aquinas rejects the possibility that a substance participates in the *esse* which is intrinsic to it in the way a species participates in a genus because what is included in the essence or nature of a substance is part of its definition and *ens* is neither a genus nor a specific difference.[90] *Esse* is not participated in by a substance the same way a genus is participated in by a species because a genus is divided into its species by specific differences which are external to the essence of the genus. Through the adding of a specific difference which is external to the essence of the genus to which it is added, a genus is contracted and determined to a species of it. The division of *ens* as a genus into different species of it requires that there are specific differences which are external to the essence of the genus *ens* which is *esse*. But it is impossible that there are such specific

88 Cf. *The Metaphysical Thought of Thomas Aquinas*, 105.

89 Cf. *DEE*, cap. 2: "Sic ergo patet quod essentia hominis et essentia Socratis non differt nisi secundum signatum et non signatum; unde Commentator dicit super VII Metaphysicae: Socrates nihil aliud est quam animalitas et rationalitas, quae sunt quiditas eius. Sic etiam essentia generis et speciei secundum signatum et non signatum differunt, quamvis alius modus designationis sit utrobique: quia designatio individui respectu speciei est per materiam determinatam dimensionibus, designatio autem speciei respectu generis est per differentiam constitutivam quae ex forma rei sumitur. Haec autem determinatio vel designatio quae est in specie respectu generis, non est per aliquid in essentia speciei exsistens quod nullo modo in essentia generis sit; immo quicquid est in specie est etiam in genere ut non determinatum. Si enim animal non esset totum quod est homo sed pars eius, non praedicaretur de eo, cum nulla pars integralis de suo toto praedicetur. [...] Et quia, ut dictum est, natura speciei est indeterminata respectu individui sicut natura generis respectu speciei: inde est quod, sicut id quod est genus prout praedicabatur de specie implicabat in sua significatione, quamvis indistincte, totum quod determinate est in specie, ita etiam et id quod est species secundum quod praedicatur de individuo oportet quod significet totum id quod est essentialiter in individuo, licet indistincte. Et hoc modo essentia speciei significatur nomine hominis, unde homo de Socrate praedicatur."

90 Cf. *QQ*, II, q. 2, a. 1, corp.: "Sed sciendum est quod aliquid participatur dupliciter. Uno modo quasi existens de substantia participantis, sicut genus participatur a specie; hoc autem modo esse non participatur a creatura: id enim est de substantia rei quod cadit in eius definitione, ens autem non ponitur in definitione creaturae, quia nec est genus neque est differentia."

differences because outside the essence of *ens*, i.e., outside *esse*, is nothing and nothing cannot contract and determine the genus *ens* to a species of it.[91]

Ens is not included as a specific difference in the definition, essence, or nature of a substance which participates in *esse* either because *ens* does not determine a substance in any quidditative way, i.e., as that which it is, as a specific difference does. To sum up: *Esse* is not participated in by a substance like a genus is participated in by a species of it because *ens* is not part of the definition, essence, or nature of the participating substance.

Before I get to the possibility that *esse* is participated in as something which is not included in the essence or nature of that which participates, another case must be considered where something is participated in as something which is included in the essence or nature of that which participates. The question is whether *esse* is participated in by a substance as form is participated in by matter.[92] In Aquinas's view, this case of participation is different from the case of the

91 Cf. *In Meta*, V, lect. 9, 889: "Sciendum est enim quod ens non potest hoc modo contrahi ad aliquid determinatum, sicut genus contrahitur ad species per differentias. Nam differentia, cum non participet genus, est extra essentiam generis. Nihil autem posset esse extra essentiam entis, quod per additionem ad ens aliquam speciem entis constituat: nam quod est extra ens, nihil est, et differentia esse non potest. Unde in tertio huius probavit Philosophus, quod ens, genus esse non potest." But could not various privations serve as such specific differences which determine and contract *ens* as a genus to various species? For example, it could be argued that God is an *ens* which lacks nothing of *esse*, creatures are *entia* which lack the *esse* of the divine attributes (e.g., omnipotence, omniscience etc.), human souls are *entia* which lack the *esse* of existing *in se*, non-human animals are *entia* which lack the *esse* of rationality, plants are *entia* which lack the *esse* of sentience, inanimate objects are *entia* which are lack the *esse* of life, etc. I think privations cannot serve as specific differences which determine and contract *ens* as a genus to various species because specific differences are forms which determine a genus in a formal way. But a privation is not a form but a privation of a form. Thus, a privation logically and ontologically presupposes that something is already formally constituted as that which it is. Furthermore, the lack of a form—and consequently a lack of *esse*—is only a privation if a substance is supposed to have that form and *esse* in virtue of what it is essentially but fails to have it for some contingent reason. For example, the division of the genus *sensitive animate corporeal substance*, i.e, the division of the genus *animal* into the species *rational animal* and *non-rational animal* is not a division by privation. The lack of rationality of non-rational animals is not a privation because rationality is not part of the essence or nature of non-rational animals. There is nothing wrong or deficient with non-rational animals which are non-rational in contrast to animals that should be able to see in virtue of what they are, but which are blind for some contingent reason. I am thankful to Jason T. Eberl for raising this objection and for making me aware of the need to explain why this is a dead-end route.

92 Cf. *In BDH*, lect. 2, 24: "Similiter etiam subiectum participat accidens et materia formam, quia forma substantialis vel accidentalis, quae de sui ratione communis est, determinatur ad hoc vel illud subiectum." For a more detailed analysis of Aquinas's view on the second mode

participation of an individual in a species and a species in a genus because it is a case of a real or ontological participation, i.e., a case where something is not only logically but metaphysically composed.[93]

However, *esse* is not participated by a substance as form is participated in by matter because form which is participated in by matter is included in the essence or nature of the resulting material substance, but a substance's participating in *esse* does not result in a substance whose essence or nature includes *esse*. In other words, a substance does not participate in *esse* in such a way that a substance results whose essence or nature is not simple but composed of form and *esse*.[94] Rather, *esse* is participated in as something which is outside the essence or nature of the participating substance so that a substance results which is metaphysically composed of a simple or complex essence or nature and *esse*.[95] To sum up: According to Aquinas, *esse* is not participated in as something which is included in the essence or nature of that which participates.

This leaves us with the second option: *Esse* is participated in as something which is not included in the essence or nature of that which participates. And this is indeed Aquinas's view: *Esse* is participated in by a substance as something which is not included in its essence or nature.[96] But now the decisive ques-

---

of participation, see Daniel De Haan, "Aquinas on *Actus Essendi* and the Second Mode of Participation," *The Thomist* 82, no. 4 (2018).

93 I think this is the reason why Aquinas lumps together this way of participation and the way an accidental form is participated in by a subject. Nevertheless, the way that a genus is participated in by a species (and a species by an individual) and the way a form is participated in by matter have in common that they are cases where something is participated in as something which is included in the essence or nature of that which participates. For this reason, I treat the latter two ways of participation together in opposition to the case where an accidental form is participated in by a subject, which is a case where something is participated in as something which is not included in the essence or nature of that which participates.

94 Cf. *QQ*, II, q. 2, a. 1, ad. 1: "Ad primum ergo dicendum quod aliquando ex hiis quae simul iunguntur relinquitur aliqua res tertia, sicuti ex anima et corpore constituitur humanitas, quae est homo, unde homo componitur ex anima et corpore; aliquando autem ex hiis quae simul iunguntur non resultat res tertia, sed resultat quaedam ratio composita, sicut ratio hominis albi resolvitur in rationem hominis et in rationem albi, et in talibus aliquid componitur ex se ipso et alio, sicut album componitur ex eo quod est album et ex albedine."

95 Cf. *QQ*, II, q. 2, a. 1, corp.: "Sic ergo in angelo est compositio ex essentia et esse, non tamen est compositio sicut ex partibus substantiae, sed sicut ex substantia et eo quod adhaeret substantiae."

96 Cf. *QQ*, II, q. 2, a. 1, corp.: "Id enim est de substantia rei quod cadit in eius definitione, ens autem non ponitur in definitione creaturae, quia nec est genus neque est differentia. Unde participatur sicut aliquid non existens de essentia rei, et ideo alia quaestio an est et quid est; unde, cum omne quod est praeter essentiam rei dicatur accidens, esse, quod pertinet ad quaestionem an est, est accidens. Et ideo Commentator dicit in V Metaphysicae quod ista propositio: Sortes

tion is whether this view implies that *esse* is an accident or accidental form of a substance which has it, so that a substance is accidentally related to its *esse*.

Aquinas negates this: A substance does not have *esse* as an accident or accidental form, and a substance is not accidentally related to the *esse* in virtue of which it is actual as that which it is essentially.[97] What needs to be explained then is how a substance can participate in *esse* as something which is not included in its essence or nature without it being the case that a substance has *esse* as an accidental form and is accidentally related to it. In other words, what needs to be explained is how the participation of a substance in *esse* distinguishes itself from the participation of a substance or subject in an accidental form it has.[98]

To explain how it is possible that *esse* is participated in by a substance without it being the case that *esse* is an accidental form of the participating substance, Aquinas introduces a further distinction between two ways that something is participated in as something which is not included in the essence or nature of that which participates. Either something is participated in as something which is outside the essence of that which participates but nevertheless determines the essence of that which participates, or something is participated in as something which is outside the essence of that which participates and does not determine the essence of that which participates.[99]

---

est, est de accidentali praedicato, secundum quod importat entitatem rei vel veritatem propositionis, sedverum est quod hoc nomen ens, secundum quod importat rem cui competit huiusmodi esse, sic significat essentiam rei, et dividitur per decem genera."; *SCG*, II, cap. 53: "In quocumque enim inveniuntur aliqua duo quorum unum est complementum alterius, proportio unius eorum ad alterum est sicut proportio potentiae ad actum: nihil enim completur nisi per proprium actum. In substantia autem intellectuali creata inveniuntur duo: scilicet substantia ipsa; et esse eius, quod non est ipsa substantia, ut ostensum est. Ipsum autem esse est complementum substantiae existentis: unumquodque enim actu est per hoc quod esse habet. Relinquitur igitur quod in qualibet praedictarum substantiarum sit compositio actus et potentiae."

97 Cf. *In Meta*, IV, lect. 2, 558: "Sed in primo quidem non videtur dixisse recte. Esse enim rei quamvis sit aliud ab eius essentia, non tamen est intelligendum quod sit aliquod superadditum ad modum accidentis, sed quasi constituitur per principia essentiae."; *QQ*, XII, q. 4, corp.: "Et sic dico quod esse substantiale rei non est accidens, sed actualitas cuiuslibet formae existentis, sive sine materia sive cum materia."; *QQ*, II, q. 2, a.1, ad 2: "Ad secundum dicendum quod esse est accidens, non quasi per accidens se habens, sed quasi actualitas cuiuslibet substantiae."

98 Cf. *In BDH*, lect. 2, 24: "Similiter etiam subiectum participat accidens et materia formam, quia forma substantialis vel accidentalis, quae de sui ratione communis est, determinatur ad hoc vel illud subiectum."

99 Cf. *QQ*, II, q. 2, a.2, ad 1: "Hoc tamen est aliter et aliter in utrisque. Dupliciter enim aliquid accipitur ut accidens praeter rationem rei. Uno modo, quia non cadit in definitione significante essentiam rei, sed tamen est designativum vel determinativum alicuius essentialium principio-

Let us begin with the first case. Aquinas presents two examples to illustrate this way or mode of participation. The first example is the specific difference *rational* which is participated in by the genus *animal*. *Rational* is participated in by *animal* as something which is not included in the definition or essence of *animal*.[100] If *rational* were included in the definition or essence of *animal*, it would be impossible that there are non-rational animals. Nevertheless, the specific difference *rational* which is participated in by the genus *animal* determines and contracts the essence of the participating genus *animal* to the species *rational animal*. As we have seen above, it is important to note that the essence of the genus *animal* and the essence of the species *rational animal* only differ as the undetermined and the determined differ.[101] A rational animal is essentially an animal. Consequently, the participatory relationship between a specific difference, a genus, and the resulting species can be described as a relation between a determinant, a determinable, and a determinate. A species is a determinate which results from a determinable genus being determined by participating in a specific difference which is outside the essence of the participating determinable genus.[102]

---

rum. [...] Alio modo accidit aliquid alicui, quia nec est in eius definitione, nec est determinativum alicuius essentialium principiorum."

100 Cf. *QQ*, II, q. 2, a.2, ad 1: "Uno modo, quia non cadit in definitione significante essentiam rei, sed tamen est designativum vel determinativum alicuius essentialium principiorum, sicut rationale accidit animali, utpote praeter definitionem eius existens, et tamen est determinativum essentiae animalis, unde fit essentiale homini et de ratione eius existens."

101 Cf. *DEE*, cap. 2: "Sic etiam essentia generis et speciei secundum signatum et non signatum differunt, quamvis alius modus designationis sit utrobique: quia designatio individui respectu speciei est per materiam determinatam dimensionibus, designatio autem speciei respectu generis est per differentiam constitutivam quae ex forma rei sumitur. Haec autem determinatio vel designatio quae est in specie respectu generis, non est per aliquid in essentia speciei exsistens quod nullo modo in essentia generis sit; immo quicquid est in specie est etiam in genere ut non determinatum. Si enim animal non esset totum quod est homo sed pars eius, non praedicaretur de eo, cum nulla pars integralis de suo toto praedicetur. [...] Et quia, ut dictum est, natura speciei est indeterminata respectu individui sicut natura generis respectu speciei: inde est quod, sicut id quod est genus prout praedicabatur de specie implicabat in sua significatione, quamvis indistincte, totum quod determinate est in specie, ita etiam et id quod est species secundum quod praedicatur de individuo oportet quod significet totum id quod est essentialiter in individuo, licet indistincte."

102 Cf. *SCG*, II, cap. 95: "Ratio igitur determinatae speciei consistit in hoc quod natura communis in determinato gradu entis collocatur. Et quia in rebus ex materia et forma compositis forma est quasi terminus, id autem quod terminatur per eam est materia vel materiale: oportet quod ratio generis sumatur ex materiali, differentia vero specifica ex formali. Et ideo ex differentia et genere fit unum sicut ex materia et forma. Et sicut una et eadem est natura quae ex materia

Esse is not participated in as a specific difference is participated in by a genus because the participation of a substance in *esse* does not result in an *ens* by essence, i.e., an *ens* which has *ens* included in its definition or essence for the reasons we have seen above.[103] *Esse* is not participated in by a substance as a specific difference is participated in by a genus because *esse* does not determine the essence of the participating substance as a specific difference does. A substance's *esse* does not place a substance into a species by adding a specific difference to its genus. *Esse* does not determine a substance in any quidditative way, i.e., as what it is essentially, as a specific difference does.

Aquinas's second example, which he uses to illustrate the case that something is participated in as something which is outside the essence of that which participates but nevertheless determines the essence of that which participates, is restricted to material substances. Only in the case of material substances are there individual members of a species which participate in the common essence or nature of that species. The idea is this: The essence of a material substance is metaphysically composed of form and matter.

The matter which is part of the essence of a material substance is non-designated matter, i.e., the *type* of matter out of which the individual members of the species are made.[104] For example, the non-designated matter which is part of the definition or essence of a human being is flesh and bones but not *this* flesh and *these* bones.[105] Thus, what is part of the essence of a human being is matter which is undetermined with respect to the dimensions which locate it in three-dimensional space.[106] Yet any individual human being which exists does not exist as a human being which has non-designated matter but designat-

---

et forma constituitur, ita differentia non addit quandam extraneam naturam super genus, sed est quaedam determinatio ipsius naturae generis."

103 Cf. *QQ*, II, q. 2, a.1, corp.: "Sed sciendum est quod aliquid participatur dupliciter. Uno modo quasi existens de substantia participantis, sicut genus participatur a specie; hoc autem modo esse non participatur a creatura: id enim est de substantia rei quod cadit in eius definitione, ens autem non ponitur in definitione creaturae, quia nec est genus neque est differentia."

104 Cf. Kerr, *Aquinas's Way to God*, 43.

105 Cf. *DEE*, cap. 2: "In diffinitione autem hominis ponitur materia non signata: non enim in diffinitione hominis ponitur hoc os et haec caro, sed os et caro absolute quae sunt materia hominis non signata."

106 Cf. *DEE*, cap. 2: "Sic etiam essentia generis et speciei secundum signatum et non signatum differunt, quamvis alius modus designationis sit utrobique: quia designatio individui respectu speciei est per materiam determinatam dimensionibus, designatio autem speciei respectu generis est per differentiam constitutivam quae ex forma rei sumitur."

ed matter, i.e., matter which is determined with respect to the dimensions which locate it in three-dimensional space.[107]

According to Aquinas, the dimensions which determine non-designated matter to be designated matter are accidental forms which belong to the genus *quantity*.[108] Consequently, in the case of a material substance, it participates in something which is outside its essence, but which nevertheless determines its

---

107 Cf. *DEE*, cap. 2: "Et dico materiam signatam quae sub determinatis dimensionibus consideratur."

108 Cf. *DEE*, cap. 2: "Corpus enim secundum quod est in genere substantiae dicitur ex eo quod habet talem naturam ut in eo possint designari tres dimensiones; ipsae enim tres dimensiones designatae sunt corpus quod est in genere quantitates."; *In BDT*, q. 4, a. 2, corp.: "Materia autem non est divisibilis nisi per quantitatem; unde Philosophus dicit in I Physicorum quod subtracta quantitate remanebit substantia indivisibilis; et ideo materia efficitur haec et signata secundum quod subest dimensionibus. Dimensiones autem istae possunt dupliciter considerari. Uno modo secundum earum terminationem; et dico eas terminari secundum determinatam mensuram et figuram, et sic ut entia perfecta collocantur in genere quantitates." It is important to note that, in Aquinas's view, the dimensions or accidental forms which determine non-designated matter to be designated matter are not responsible for the individuation of a material substance. Rather, the principle of individuation is matter under undetermined dimensions, i.e., matter which is extended in three-dimensional space but where the degree of extension in any dimension is not specified. I understand Aquinas's position as saying that it is matter and not the determined dimensions or accidental forms in the genus of *quantity* which individuates a material substance because matter is the subject of the undetermined dimensions or imperfect accidents in the genus of *quantity*, cf., ibid.: "Et sic non possunt esse principium individuationis, quia cum talis terminatio dimensionum varietur frequenter circa individuum, sequeretur quod individuum non remaneret semper idem numero. Alio modo possunt considerari sine ista determinatione, in natura dimensionis tantum, quamvis numquam sine aliqua determinatione esse possint, sicut nec natura coloris sine determinatione albi et nigri; et sic collocantur in genere quantitatis ut imperfectum, et ex his dimensionibus indeterminatis materia efficitur haec materia signata, et sic individuat formam. Et sic ex materia causatur diversitas secundum numerum in eadem specie. Unde patet quod materia secundum se accepta nec est principium diversitatis secundum speciem nec secundum numerum; sed sicut est principium diversitatis secundum genus prout subest formae communi, ita est principium diversitatis secundum numerum prout subest dimensionibus interminatis." It should be noted that there is an ongoing debate about Aquinas's view on individuation among Thomistic scholars. Since nothing important in my argument hinges on this issue, I do not enter this debate. For similar interpretations of Aquinas's view on individuation, see Brown, *Aquinas and the Ship of Theseus*, 124–130; Skrzypek, "Accidental Forms as Metaphysical Parts."; Stump, *Aquinas*, 47–50; Wippel, *The Metaphysical Thought of Thomas Aquinas*, 357–373. Good overviews of the more recent development of the debate and references to competing interpretations of Aquinas's view on individuation can be found in Adam Wood, *Thomas Aquinas on the Immateriality of the Human Intellect* (Washington, D.C.: The Catholic University of America Press, 2020), 58–64; Gaston G. LeNotre, "Determinate and Indeterminate Dimensions: Does Thomas Aquinas Change His Mind on Individuation?," *American Catholic Philosophical Quarterly* 94, no. 4 (2020).

essence.[109] Accidental forms of the genus *quantity* are participated in as something not included in the essence of the participating material substance, but these accidental forms nevertheless determine the essence of the participating material substance. They determine the non-designated matter which is part of the essence of the participating substance to be the designated matter which the individual has in actuality.

Again, the participatory relationship between an individual material substance whose matter is designated, its essence or nature which includes non-designated matter, and accidental forms of the genus *quantity* which are outside of the essence of the material substance can be described as a relation between a determinant, a determinable, and a determinate. An individual material substance with designated matter is the determinate which results from its determinable essence or nature—which is determinable because it contains matter which is undetermined with respect to the dimensions which locate it in three-dimensional space—being determined by participating in accidental forms of the genus *quantity.*

But *esse* is not participated in by a substance like accidental forms of the genus *quantity* are participated in by the non-designated matter of a substance for the simple reason that only the essence of a material substance contains matter which is determinable through accidental forms of the genus *quantity* which are outside the essence of the participating material substance. Thus, this cannot be the way *esse* is participated in by a substance because in the case of an immaterial substance there is nothing outside its essence which could determine its essence. Immaterial substances like angels are substantial forms which are not received in matter and are therefore fully individualized by their substantial form.[110]

109 Cf. *QQ*, II, q. 2, a. 2, ad 1: "Hiis igitur quae sunt composta ex materia et forma, accidit aliquid praeter rationem speciei existens utroque modo. Cum enim de ratione speciei humanae sit quod componatur ex anima et corpore, determinatio corporis et animae est praeter rationem speciei, et accidit homini in quantum est homo quod sit ex hac anima et ex hoc corpore; sed hoc convenit per se huic homini, de cuius ratione esset, si definiretur, quod esset ex hac anima et ex hoc corpore, sicut de ratione hominis communis est quod sit ex anima et corpore. Accidunt etiam compositis ex materia et forma praeter rationem speciei multa alia quae non sunt determinativa essentialium principiorum." Nevertheless, they possess accidents by means of which they have a multiplicity of accidental *esse.* I am grateful to Jeremy Skrzypek for helping me to see that this point should be made explicit.

110 Cf. *QQ*, II, q. 2, a. 2, ad 1: "Substanciis vero inmaterialibus creatis accidunt quidem aliqua praeter rationem speciei quae non sunt determinativa essentialium principiorum, ut dictum est; non tamen accidunt eis aliqua quae sunt determinativa essentiae speciei, quia ipsa natura spe-

Furthermore, if *esse* were participated in by material substances in this way, it would follow that *esse* is an accidental form of a material substance and that a material substance is accidentally related to its *esse*. But this cannot be true. A material substance can lose the specific accidental forms of the genus *quantity* by which it is extended in a certain way in three-dimensional space and gain others by which it is extended in three-dimensional space in a different way. A material substance is not accidentally related to the *esse* it has and in virtue of which it is actual as that which it is essentially because it cannot lose its *esse* and continue to exist. Thus, *esse* is not participated in by a substance like accidental forms of the genus *quantity* are participated in by the non-designated matter of a material substance.

As far as I can see, these two examples exhaust the possibilities that something is participated in as something which is outside the essence of that which participates but nevertheless determines the essence of that which participates. So, we can discard the option that *esse* is participated in as something which is outside the essence of that which participates but nevertheless determines the essence of that which participates. Consequently, we are left with the option that something is participated in as something which is outside the essence of that which participates and does not determine the essence of that which participates. Aquinas's example to illustrate this case is the accidental form of whiteness which is participated in by a human being as something which is not included in its essence or nature, and which does not determine its essence or nature.[111]

In Aquinas's view, *esse* is participated in by a substance in a *similar* way, i.e., *esse* is participated in by a substance as something which is not included in the essence of the participating substance, and which does not determine its essence or nature.[112] Nevertheless, it is obvious that this cannot be *exactly*

ciei non individuatur per materiam, sed per se ipsam, ex hoc quod talis forma non est nata recipi in aliqua materia; unde per se ipsam est non multiplicabilis neque praedicabilis de pluribus."

111 Cf. *QQ*, II, q. 2, a. 2, ad 1: "Alio modo accidit aliquid alicui, quia nec est in eius definitione, nec est determinativum alicuius essentialium principiorum, sicut albedo accidit homini."

112 Cf. *QQ*, II, q. 2, a. 2, ad 1: "Substanciis vero inmaterialibus creatis accidunt quidem aliqua praeter rationem speciei quae non sunt determinativa essentialium principiorum, ut dictum est. [...] Sed quia non est suum esse, accidit ei aliquid praeter rationem speciei, scilicet ipsum esse, et alia quaedam quae attribuuntur suppositio et non naturae. Propter quod suppositum in eis non est omnino idem cum natura."; ibid., ad 2: "Ad secundum dicendum quod non omne quod accidit alicui praeter rationem speciei, est determinativum essentiae ipsius, ut oporteat illud poni in ratione eius, sicut dictum est; et ideo, licet ipsum esse non sit de ratione suppositi, quia tamen pertinet ad suppositum et non est de ratione naturae, manifestum est quod suppositum et natura non sunt omnino idem in quibuscumque res non est suum esse."

the way that *esse* is participated in by a substance. If it were, Aquinas's answer to the nature question would collapse into the view that a substance has *esse* as an accidental form and that it is accidentally related to the *esse* it has. This is not Aquinas's view. As we have seen above, Aquinas rejects the view that *esse* is an accidental form of a substance. It is not difficult to see why.

First, if *esse* were a form in which the substance participates as something which is not included in its essence or nature, *esse* would be a form which the substance has in addition to the essence or form in virtue of which it is what it is essentially and would be accidentally related to this additional form. Thus, a substance could lose this additional form or could not participate in this additional form and continue to exit. But a substance cannot lose the *esse* it has as that which it is or cease to participate in it and continue to exist.[113] Consequently, *esse* is not participated in by a substance as a form which the substance has in addition to its essence.

Second, if *esse* were a form which the substance has in addition to its essence, *esse* would determine the substance in an accidental formal respect. But it does not. *Esse* does not determine its substance with respect to its quality, quantity, relation etc. or any other category of accidents. *Esse* does not fall under any category of accidents. Thus, if one wants to stick with the claim that *esse* is an accidental form of a substance, one must claim that *esse* is a kind of strange *sui generis* accidental form which does not fit into any category of accidents and which does not determine its substance in any accidental formal respect. However, if *esse* does not determine or modify its substance in any accidental formal respect, it is unintelligible why we should consider it to be an accidental form at all.

Third, if *esse* were participated in by a substance as a form which the substance has in addition to its essence, the whole explanation of what it is for a substance to exist would become viciously circular. A substance can only participate in *esse* as an accidental form which is outside its essence if the substance has *esse* and is actual as that which it is essentially. Thus, the participation of a substance in *esse* as an accidental form does not explain its having *esse* but rather presupposes it. Put otherwise, *esse* as an accidental form can only inhere in a substance which participates in it if this substance is actual as that which it is.

It follows that Aquinas must introduce a further distinction which allows him to distinguish the way *esse* is participated in by a substance from the way

113 Cf. *QDP*, q. 5, a. 1, ad 10: "Ad decimum dicendum, quod etiam praedicta actione cessante, forma deficeret, unde non posset esse essendi principium."

an accidental form is participated in by a substance. These two modes are alike in two respects: first, what is participated in by a substance is extrinsic to its essence or nature. Second, what is participated in does not determine the substance's essence or nature.

What distinguishes the two modes or ways of participation? The decisive difference between the two ways of participation is that only *esse* is participated in by a substance as something which determines its essence or nature in a certain way. Namely, *esse* is participated in by a substance as something which is outside its essence or nature but which determines the essence or nature of the participating substance and consequently the substance as that which it is essentially with respect to its actuality.[114] An accidental form is not participated in by a substance in this way. Rather, an accidental form can be participated in by a substance only if the participating substance is already actual as that which it is essentially. Thus, the participation of a substance in an accidental form it has presupposes its participation in *esse*, i.e., the participation of a substance's essence or form in *esse*.

It is important to note that this way of distinguishing between the way or mode *esse* is participated in by a substance from the way or mode an accidental form is participated in by a substance does not make Aquinas's view inconsistent. For example, one could try to object that the claim that *esse* is participated in as something which determines the essence or nature of the participating substance and consequently the substance as what it is essentially with respect to its

114 Cf. *QQ*, II, q. 2, a. 1, ad. 2: "Ad secundum dicendum quod esse est accidens, non quasi per accidens se habens, sed quasi actualitas cuiuslibet substantiae."; *QQ*, XII, q. 4, corp.: "Unde esse est completivum omnis formae, quia per hoc completur quod habet esse, et habet esse cum est actu; et sic nulla forma est nisi per esse. Et sic dico quod esse substantiale rei non est accidens, sed actualitas cuiuslibet formae existentis, sive sine materia sive cum materia."; *SCG*, II, cap. 53: "Omne participans aliquid comparatur ad ipsum quod participatur ut potentia ad actum: per id enim quod participatur fit participans actu tale."; *QDP*, q. 7, a. 2, ad 9: "Ad nonum dicendum, quod hoc quod dico esse est inter omnia perfectissimum: quod ex hoc patet quia actus est semper perfectio potentia. Quaelibet autem forma signata non intelligitur in actu nisi per hoc quod esse ponitur. Nam humanitas vel igneitas potest considerari ut in potentia materiae existens, vel ut in virtute agentis, aut etiam ut in intellectu: sed hoc quod habet esse, efficitur actu existens. Unde patet quod hoc quod dico esse est actualitas omnium actuum, et propter hoc est perfectio omnium perfectionum."; *ST*, I, q. 4, a. 1, ad. 3: "Ad tertium dicendum quod ipsum esse est perfectissimum omnium, comparatur enim ad omnia ut actus. Nihil enim habet actualitatem, nisi inquantum est. Unde ipsum esse est actualitas omnium rerum, et etiam ipsarum formarum. Unde non comparatur ad alia sicut recipiens ad receptum, sed magis sicut receptum ad recipiens. Cum enim dico esse hominis, vel equi, vel cuiuscumque alterius, ipsum esse consideratur ut formale et receptum, non autem ut illud cui competit esse."

actuality conflicts with his claim that *esse* is participated in as something which does not determine the essence or nature of the participating substance.

There is no inconsistency because the latter claim just excludes the possibility that *esse* determines the essence or nature of the participating substance as a specific difference determines a genus or as accidental forms of the genus *quantity* determine the non-designated matter which is part of the essence of a material substance. In short, what has been excluded by Aquinas is the option that *esse* determines the essence or nature of the participating substance in any quidditative way. The claim that *esse* determines the essence or nature of the participating substance with respect to its actuality does not imply that *esse* determines the participating substance in any quidditative respect. On the contrary, the participation of a substance in *esse* does not determine in any way what a substance is, i.e., a reference to the *esse* of a substance does not help in any way to answer the question 'What is it?'.[115] It is therefore perfectly consistent to claim that *esse* is participated in as something which does not determine the essence or nature of the participating substance in any quidditative way *and* as something which determines the essence or form of the participating substance and consequently the substance as that which it is essentially with respect to its actuality.

We can conclude that Aquinas's view can satisfyingly explain how *esse* is participated in as something which is not included in the essence or nature of the participating substance without it being the case that *esse* is an accidental form of that substance so that the substance is accidentally related to the *esse* it has.

Aquinas can solve this problem with his view that *esse* is participated in by the essence or form of a substance.[116] This view is able to elegantly reconcile two claims which at first glance seem to contradict each other, namely, the claim that a substance is an *ens* by participation and the claim that a substance has substantial *esse*, i.e., that it subsists as that which it is. A substance is an *ens* by participation and not an *ens* by essence because *esse* is not included in its essence. Consequently, no substance is essentially an *ens*, i.e., it does not have *esse* necessarily. The fact that a substance is an *ens* by participation accounts for the contingency of its existence.

---

115 Cf. *QQ*, II, q. 2, a. 1, corp.: "Unde participatur sicut aliquid non existens de essentia rei, et ideo alia quaestio an est et quid est; unde, cum omne quod est praeter essentiam rei dicatur accidens, esse, quod pertinet ad quaestionem an est, est accidens."

116 Cf. *QDSC*, a. 1, corp.: "Omne igitur quod est post primum ens, cum non sit suum esse, habet esse in aliquo receptum, per quod ipsum esse contrahitur; et sic in quolibet creato aliud est natura rei quae participat esse, et aliud ipsum esse participatum."

Nevertheless, the fact that a substance exists only contingently does not imply that a substance is accidentally related to its *esse*. A substance has substantial *esse*, i.e., it subsists as that which it is, precisely because it is its essence or form which participates in *esse*. By participating in *esse*, the essence or form of a substance and consequently the substance itself as that which it is essentially is determined with respect to its actuality. In other words, if the essence or form of a substance ceases to participate in *esse*, the whole substance ceases to exist. Thus, according to Aquinas's view, it makes perfect sense to claim that a substance is an *ens* by participation which nevertheless has substantial *esse* and subsists as that which it is essentially.

Let us recapitulate what we have seen in this section. I have shown in detail how Aquinas makes use of the idea of participation to explain how a substance is related to its essence or form and its *esse* in such a way that it has *esse* by means of its essence or form. I called this the 'second stage' of Aquinas's explanation of what it is for a substance to exist because with the help of the idea of participation Aquinas can explain in a precise way how a substance has *esse* by means of its essence or form. A substance has *esse* by means of its essence or form in virtue of the participation of its essence or form in *esse*. Aquinas explains how *esse* is participated in by the essence or form of a substance by carefully distinguishing it from several other ways that something is participated in by something else.

First, the *esse* which is participated in by the essence or form of a substance is intrinsic to the substance. Second, *esse* is participated in as something which is intrinsic to the substance but outside of the essence or nature of the participating substance. Third, *esse* is participated in as something which is outside of the essence or nature of the participating substance and as something which does not determine the essence or nature of a substance in any quidditative way, i.e., as that which it is essentially. The way *esse* is participated in by a substance has this characteristic in common with the way an accidental form is participated in by a substance. Finally, what distinguishes the way *esse* is participated in by a substance from the way an accidental form is participated in by a substance is that only *esse* determines the essence or nature and consequently the substance as what it is essentially with respect to its actuality. *Esse* determines a substance to be in actuality as what it is essentially precisely because it is the essence or form of a substance which participates in *esse* and not any accidental form which the substance has in addition to its essence or form.

In this way, Aquinas's view can explain how *esse* is participated in by a substance as something which is outside of its essence or nature without it being the case that *esse* is an accidental form of that substance such that a substance is accidentally related to the *esse* in virtue of which it is actual as what it is essentially. *Esse* is also participated in by an accidental form a substance has but this participation in *esse* only determines a substance to be in actuality in an accidental respect precisely because it is only an accidental form of that substance which participates in *esse*.

6.3 Participation in *Esse* Results from a Metaphysical Composition of a Determinable and a Determinant

Up to this point, I have reconstructed two stages of Aquinas's explanation of what it is for a substance to exist. Aquinas's first and preliminary answer to the nature question is that for a substance to exist is to have *esse* by means of its essence or form (see 6.1). This answer is not wrong but still unsatisfactory because it neither clarifies sufficiently how this 'having' works, nor does it explain how a substance is related to its essence or form and its *esse*. In particular, at this stage it remains unexplained how a substance can have *esse* without it being the case that *esse* is an accident of it such that a substance is accidentally related to its *esse*. Yet, as I have shown in the previous section (see 6.2), Aquinas does not content himself with stating that for a substance to exist is to have *esse* by means of its essence or form. What the first stage leaves unexplained is addressed at the second stage of his explanation of what it is for a substance to exist.

At this second stage, Aquinas introduces the idea of participation to formulate a more satisfactory answer to the nature question. He carefully distinguishes the way *esse* is participated in by a substance from several other ways that something is participated in. This puts him into a position to explain how a substance has *esse* by means of its essence or form. *Esse* is participated in as something:
1. which is intrinsic to the participating substance
2. which is outside of the essence or nature of the participating substance
3. which does not determine the essence or nature of a substance in any quidditative way
4. which does not determine a substance in any accidental formal respect
5. which does determine the essence or form of a substance and consequently the substance itself as that which it is essentially with respect to its actuality.

Thus, a substance's having *esse* is explained in terms of participation. Furthermore, Aquinas's distinction between the way an accidental form is participated in by a substance (4) and the way *esse* is participated in by a substance (5) allows him to clarify how a substance is related to its *esse* and how a substance can participate in *esse* without it being the case that its *esse* is one of its accidents, or that a substance is accidentally related to the *esse* in which it participates. A substance does not participate in *esse* as it participates in an accidental form it has in addition to its essence or form. Rather, what participates in *esse* is the essence or form of the substance.[117] *Esse* is not an accidental form which the substance has in addition to its essence or form but the actuality of its essence or form which its essence or form has in virtue of its participation in *esse*.[118] Thus, a substance is related to its *esse* as that by which its essence or form—and consequently the substance itself taken as what it is essentially—is actual. In other words, *esse* determines the essence or form of a substance with respect to its actuality by means of which the substance has *esse* and is actual as what it is essentially.

Due to this mediatory role of the essence or form of a substance, it is possible to conceive the relationship between a substance and its essence or form from two different perspectives. Considered as that which participates in *esse* in abstraction from the *esse* in which it participates, the essence or form of a substance is a principle of potentiality of a substance because it is merely that by which the substance participates in *esse* and receives *esse*.[119] Considered in abstraction from *esse*, any essence or form of a substance—even a simple essence or form like that of an immaterial substance or angel—is just a principle of potentiality of that substance because it stands in potentiality to the *esse* in which it participates.[120]

Nevertheless, it is important to note that this potentiality of the essence or form of a substance as that which participates in *esse* cannot be *separated* from the actuality it has in virtue of its participation in *esse*.[121] In other words,

---

117 Cf. *QDSC*, a. 1, corp.: "In quolibet creato aliud est natura rei quae participat esse, et aliud ipsum esse participatum."

118 Cf. *QQ*, II, q. 2, a. 1, ad. 2: "Ad secundum dicendum quod esse est accidens, non quasi per accidens se habens, sed quasi actualitas cuiuslibet substantiae."; *QQ*, XII, q. 4, corp.: "Unde esse est completivum omnis formae, quia per hoc completur quod habet esse, et habet esse cum est actu; et sic nulla forma est nisi per esse. Et sic dico quod esse substantiale rei non est accidens, sed actualitas cuiuslibet formae existentis, sive sine materia sive cum materia."

119 Cf. *In Phy*, VIII, lect. 21, 1153: "Omnis ergo substantia quae est post primam substantiam simplicem, participat esse. Omne autem participans componitur ex participante et participato, et participans est in potentia ad participatum."

120 Cf. *QDSC*, a. 1, corp.: "Et hoc modo natura spiritualis substantiae, quae non est composita ex materia et forma, est ut potentia respectu sui esse; et sic in substantia spirituali est compositio potentiae et actus."; *In Phy*, VIII, lect. 21, 1153: "In omni ergo substantia quantumcumque simplici, post primam substantiam simplicem, est potentia essendi."

121 Cf. *QDSC*, a. 1, corp.: "Non dico autem ut potentiam separabilem ab actu, sed quam semper suus actus comitetur."

an essence or form of a substance is *really distinct* from the *esse* it participates in. However, there are no essences or forms of substances *in separation* from the *esse* they participate in. Consequently, considered together with the *esse* from which it cannot be separated, the essence or form of a substance is a principle of actuality of that substance precisely because a substance, through or by means of its essence or form, has the *esse* in virtue of which it is actual as what it is essentially.

This brings us to the third and final stage of Aquinas's explanation of what it is for a substance to exist. It could be objected that the second stage of Aquinas's answer to the nature question is still unsatisfactory because it explains something which is difficult to understand—how a substance has *esse* by means of its essence or form and how a substance has *esse* without *esse* being an accident of the substance—with something that is at least as difficult to understand, namely, how a substance *participates* in *esse*. In other words, it could be objected that Aquinas's view that the existence of a substance consists in the participation of its essence or form in *esse* is still unsatisfactory because it is notoriously difficult to understand how participation works.

I respond that the objection would have force had Aquinas failed to proceed beyond the second stage of his answer to the nature question. But there is a further layer of his explanation which addresses this possible worry. Aquinas does not only make use of the idea of participation; he also offers a precise account of how participation works. This account has the additional advantage of shedding further light on how a substance is related to its essence and its *esse*.

The third and final stage of Aquinas's explanation of what it is for a substance to exist makes something explicit which has been implicitly acknowledged at the second stage without reflecting any further on its implications (see 6.2): Metaphysical participation—in contrast to merely logical participation—requires metaphysical composition. To be more precise, metaphysical participation results from metaphysical composition.

In Aquinas's view, whenever something is participated in there has to be something in addition to or besides that which is participated in, namely, that which participates.[122] For example, if there is an illuminated body, i.e., a body which participates in light, there must be something in addition to the light which is participated in, namely, the body which participates in the light.[123] Thus, whenever something is metaphysically participated in, that which partic-

---

122 Cf. *QQ*, II, q. 2, a. 1, corp.: "Quandocumque autem aliquid praedicatur de altero per participationem, oportet ibi aliquid esse praeter id quod participatur."

123 Cf. *QQ*, II, q. 2, a. 1, corp.: "Lux enim praedicatur de corpore illuminato participative, sed, si esset aliqua lux separata, praedicaretur de ea essentialiter."

ipates has to be something which is metaphysically composed of that which participates and that which is participated in. So, in general, participation is explained by Aquinas in terms of composition, and metaphysical participation in terms of metaphysical composition.[124]

However, this is not all. Metaphysical participation requires not only that that which participates is metaphysically composed of that which participates and that which is participated in, but also that that which participates is related to that which is participated in as potentiality to actuality.[125] In other words, metaphysical participation is the result of a metaphysical composition of a determinable and a determinant.

So, according to Aquinas, any case of metaphysical participation can be explained in terms of a determinate which results from the metaphysical composition of a determinable (*quod est*) and a determinant (*quo est*). For example, the metaphysical participation of an illuminated body, i.e., the metaphysical participation of a body in light, can be explained in the following way: An illuminated body is a determinate which results from the metaphysical composition of a determinable (*quod est*) and a determinant (*quo est*). The determinable is the body considered in abstraction from the light, i.e., the body considered merely as that which has and participates in light. Considered in this way, the body has light only in potentiality. And the light is the determinant in virtue of which the body is illuminated in actuality.

To avoid a possible confusion, it is important to distinguish between that which participates in light in actuality (*quod est*)$_1$, i.e., the determinate, and that which has and participates in light in potentiality (*quod est*)$_2$, i.e., the determinable. The determinate (*quod est*)$_1$—the illuminated body—results from the metaphysical composition of a determinable (*quod est*)$_2$—the body considered in abstraction from the light, i.e., considered merely as that which has and participates in light—and a determinant (*quo est*), namely, the light in virtue of which the body (*quod est*)$_1$ is illuminated in actuality.

Now, in Aquinas's view, the participation of a substance in *esse* can be explained in an analogous way. The metaphysical participation of an *ens* or substance in *esse* can be explained in terms of a metaphysical composition of that which participates (*quod est*) and that which is participated in (*quo est*),

---

124 Cf. *In Phy*, VIII, lect. 21, 1153: "Omne autem participans componitur ex participante et participato."; *SCG*, I, cap. 22: "Omne illud quod non potest esse nisi concurrentibus pluribus, est compositum."

125 Cf. *SCG*, II, cap. 38: "Omne participans aliquid comparatur ad ipsum quod participatur ut potentia ad actum: per id enim quod participatur fit participans actu tale."

namely, *esse*, which are related as potentiality to actuality.[126] In other words, the metaphysical participation of an *ens* or substance in *esse* is the result of its metaphysical composition of a determinable (*quod est*) and a determinant (*quo est*). That any substance is an *ens* by participation can be explained in terms of a determinate which results from the metaphysical composition of a determinable (*quod est*) and a determinant (*quo est*). Since the way *esse* is participated in by an *ens* or substance is different from the way an accidental form is participated in by a substance, the corresponding metaphysical composition must be a different kind of metaphysical composition too. The participation of a substance in *esse* does not result from a metaphysical composition of a subject and an accidental form, because a substance does not participate in *esse* as an accidental form it has in addition to its essence or form.

Let us have a closer look at what kind of metaphysical composition accounts for a substance's participation in *esse*. The determinate (*quod est*)$_1$ is a substance

126 Cf. *SCG*, II, cap. 53: "Omne participans aliquid comparatur ad ipsum quod participatur ut potentia ad actum: per id enim quod participatur fit participans actu tale. Ostensum autem est supra quod solus Deus est essentialiter ens, omnia autem alia participant ipsum esse. Comparatur igitur substantia omnis creata ad suum esse sicut potentia ad actum."; *SCG*, II, cap. 54: "Unde in compositis ex materia et forma nec materia nec forma potest dici ipsum quod est, nec etiam ipsum esse. Forma tamen potest dici quo est, secundum quod est essendi principium; ipsa autem tota substantia est ipsum quod est; et ipsum esse est quo substantia denominatur ens. In substantiis autem intellectualibus, quae non sunt ex materia et forma compositae, ut ostensum est, sed in eis ipsa forma est substantia subsistens, forma est quod est, ipsum autem esse est actus et quo est. Et propter hoc in eis est unica tantum compositio actus et potentiae, quae scilicet est ex substantia et esse, quae a quibusdam dicitur ex quod est et esse; vel ex quod est et quo est."; *SCG*, I, cap. 22: "Omne illud quod non potest esse nisi concurrentibus pluribus, est compositum. Sed nulla res in qua est aliud essentia et aliud esse, potest esse nisi concurrentibus pluribus, scilicet essentia et esse. Ergo omnis res in qua est aliud essentia et aliud esse, est composita."; *QQ*, II, q. 2, a. 1, corp.: "Quandocumque autem aliquid praedicatur de altero per participationem, oportet ibi aliquid esse praeter id quod participatur, et ideo in qualibet creatura est aliud ipsa creatura quae habet esse, et ipsum esse eius. Et hoc est quod Boethius dicit in libro De hebdomadibus, quod in omni eo quod est citra Primum, aliud est esse et quod est."; *In Phy*, VIII, lect. 21, 1153: "Omnis ergo substantia quae est post primam substantiam simplicem, participat esse. Omne autem participans componitur ex participante et participato, et participans est in potentia ad participatum. In omni ergo substantia quantumcumque simplici, post primam substantiam simplicem, est potentia essendi."; *QDSC*, a. 1, corp.: "Omne igitur quod est post primum ens, cum non sit suum esse, habet esse in aliquo receptum, per quod ipsum esse contrahitur; et sic in quolibet creato aliud est natura rei quae participat esse, et aliud ipsum esse participatum. Et cum quaelibet res participet per assimilationem primum actum in quantum habet esse, necesse est quod esse participatum in unoquoque comparetur ad naturam participantem ipsum, sicut actus ad potentiam."

or *ens* by participation, i.e., an *ens* which exists *per se* and *in se.*[127] For example, a determinate is a substance like Joe Biden. The determinable (*quod est*)$_2$ is a substance's essence or form, i.e., the substance considered as what it is essentially in abstraction from the *esse* it participates in.[128] The substance as that which has and participates (*quod est*)$_2$ in *esse* is the substance considered as what it is in virtue of its essence or nature alone, i.e., in abstraction from the *esse* it has and in which it participates by means of its essence or nature. In other words, the substance as that which has and participates (*quod est*)$_2$ is the substance considered merely as the possessor of *esse* or participator in *esse* in abstraction from the *esse* it possesses and in which it participates. For example, it is Joe Biden considered not as an actual rational animal or human being, but Joe Biden considered as what he is in virtue of his essence or form and as that which participates in *esse* but considered in abstraction from the *esse* in which he participates as something outside of his essence or nature.

The determinable (*quod est*)$_2$, i.e., the substance considered as that which has or participates in abstraction from what it has or participates in, is not identical with the determinate (*quod est*)$_1$ because the latter is a metaphysical composite that has *esse* as something which is not included in its essence or nature as mere possessor of *esse* or participator in *esse* (*quod est*)$_2$.[129] The determinable is just a metaphysical part of the determinate. For example, Joe Biden is not identical with what he is in virtue of his essence or nature alone because Joe Biden has *esse* as something which is not included in his essence or nature as a human being or rational animal. Considered in abstraction from the *esse* in which he participates, Joe Biden's essence or form and consequently Joe Biden as what he is essentially is not determined to be in actuality.

---

127 Cf. *QDSC*, a. 1, ad 8.: "Sed quod est, est id quod subsistit in esse, quod quidem in substantiis corporeis est ipsum compositum ex materia et forma, in substantiis autem incorporeis est ipsa forma simplex; quo est autem, est ipsum esse participatum, quia in tantum unumquodque est, in quantum ipse esse participat. Unde et Boetius sic utitur istis vocabulis in libro de Hebdomad., dicens, quod in aliis praeter primum, non idem est quod est et esse."

128 Cf. *QDSC*, a. 1, corp.: "Et sic in quolibet creato aliud est natura rei quae participat esse, et aliud ipsum esse participatum."

129 Cf. *QQ*, II, q. 2, a. 2, corp.: "Secundum hoc igitur, cuicumque potest aliquid accidere quod non sit de ratione suae naturae, in eo differt res et quod quid est, sive suppositum et natura: nam in significatione naturae includitur solum id quod est de ratione speciei, suppositum autem non solum habet haec quae ad rationem speciei pertinent, sed etiam alia quae ei accidunt; et ideo suppositum significatur ut totum, natura autem sive quiditas ut pars formalis; ibid., ad. 2: Et ideo, licet ipsum esse non sit de ratione suppositi, quia tamen pertinet ad suppositum et non est de ratione naturae, manifestum est quod suppositum et natura non sunt omnino idem in quibuscumque res non est suum esse."

Finally, the determinant (*quo est*) is *esse*. As that which is participated in, *esse* determines the essence or nature of a substance—and consequently the substance as what it is essentially—with respect to its actuality.[130] For example, *esse* is participated in by Joe Biden as that which determines his essence or form as a human being to be in actuality, that by means of which he has *esse* as a human being and is a human being in actuality. So, *esse* is the actuality of Joe Biden's humanity by means of which he has *esse* and is actual as a human being or rational animal.

Thus, in Aquinas's view, a substance is related to its *esse* as a determinable to a determinant. Any substance is just a determinable with respect to its *esse* because it participates in *esse* as something which is outside its essence or nature. It is the *esse* which is participated in that determines the essence or nature of a substance—and consequently the substance as that which it is essentially—to be in actuality.

To sum up: A substance or *ens* by participation is a determinate (*quod est*)$_1$ which results from the metaphysical composition of a determinable *(quod est)*$_2$ and a determinant (*quo est*). The determinable *(quod est)*$_2$ is the substance considered as that which participates in *esse* in abstraction from the *esse* in which it participates, i.e., the substance considered as what it is in virtue of its essence or nature alone. The determinant (*quo est*) is *esse*, which determines the essence or nature of a substance—and consequently the substance as what it is essentially —to be in actuality. Thus, Aquinas's final answer to the nature question is that the existence of a substance consists in its metaphysical composition of a determinable and a determinant, namely, its essence or nature and its *esse*.

6.4 Is Aquinas's View Incoherent?

In the previous section, I have reconstructed the third and last stage of Aquinas's explanation of what it is for a substance to exist. Aquinas's final answer to the nature question is that the existence of a substance consists in its metaphysical composition of a determinable and a determinant, namely, its essence or nature and its *esse*. A substance is related to its *esse* as a determinable is related to a

130 Cf. *QDSC*, a. 1, ad. 8: "Quo est autem, est ipsum esse participatum, quia in tantum unumquodque est, in quantum ipse esse participat."; DEE, cap. 4: "Omne autem quod recipit aliquid ab alio est in potentia respectu illius, et hoc quod receptum est in eo est actus eius; ergo oportet quod ipsa quiditas vel forma quae est intelligentia sit in potentia respectu esse quod a Deo recipit, et illud esse receptum est per modum actus. Et ita invenitur potentia et actus in intelligentiis, non tamen forma et materia nisi aequivoce."

determinant. *Esse* determines the essence or nature of a substance—and consequently the substance itself taken as that which it is essentially—with respect to its actuality. In other words, *esse* is that which determines, and the essence or nature of a substance, i.e., the substance itself taken as what it is in virtue of its essence or nature alone in abstraction from its *esse*, is that which is determined.

A possible objection can be raised. As we have seen in the previous sections, Aquinas makes use of the ideas of participation and metaphysical composition to explain how a substance has *esse* by means of its essence or form. The existence of a substance consists in the participation of its essence or form in *esse* (see 6.2) and the participation of a substance in *esse* results from its metaphysical composition of a determinable—its essence or form—and a determinant, namely, its *esse* (see 6.3). However, if something is participated in by something else and metaphysically composed with that which participates in it, it seems that that which is participated in is restricted and limited by that which participates.

In the case of a substance which participates in *esse*, this means that *esse* which is participated in is restricted and limited by the essence or form of a substance. Consequently, it seems that the *esse* of a substance and its essence or form are related as potentiality to actuality, i.e., as a determinable and a determinant. Thus, according to the objector, Aquinas's view is incoherent because it claims that the essence or form of a substance and its *esse* are related as potentiality to actuality and as actuality to potentiality. In other words, *esse* cannot be a determinant which determines the essence or form of a substance—and consequently the substance itself as what it is essentially—with respect to its actuality, if *esse* stands itself in potentiality to the essence or form of a substance.

As far as I can see, there are three options to rebut this possible objection to Aquinas's view. The first option negates the premise that participation and metaphysical composition in general involve restriction and limitation. The second possibility concedes that participation and metaphysical composition in general involve restriction and limitation but attempts to argue that a substance's participation in *esse* is an exceptional case which does not. The third and last option concedes that *esse* which is participated in is restricted and limited by a substance's essence or form but negates that it follows from this that a substance's *esse* and its essence or form are related as potentiality to actuality so that Aquinas's view becomes incoherent.

Let us begin with the first possibility. I think it is clear that Aquinas accepts the premise that participation and metaphysical composition in general involve restriction and limitation. According to Aquinas, any kind of participation implies that that which participates in something else has only in a particular and limited way what pertains to that in which it participates in a universal and unlimited way.[131] Let us have a look at three different kinds or modes of participation which Aquinas recognizes and which we have already introduced above to explain how these kinds or modes of participation distinguish themselves from the way *esse* is participated in by a substance (see 6.2.2).

The first way something participates in something else is how a species participates in a genus and an individual in a species.[132] A genus is restricted and limited by a species which participates in it because a species which participates in the genus has only in a particular and limited way what pertains to the genus in a universal and unlimited way. For example, the species *rational animal* which participates in the genus *animal* has only in a particular and limited way what pertains to the genus *animal* in a universal and unlimited way. The species *rational animal* has only in a particular and limited way what pertains to the genus *animal* in a universal and unlimited way due to the specific difference *rational* which is added to the genus *animal* and which removes what is undetermined in the genus, namely, whether an animal has the essence or nature of a rational or non-rational animal.[133] All rational animals are animals but not all animals are rational animals. Thus, animality belongs to the species *rational animal* only in a particular and limited way. Members of the species *rational animal* only realize a particular and limited way of being an animal which as such is undetermined in the genus *animal* and therefore common to all animals.

A species is restricted and limited by an individual which participates in it because an individual which participates in a species has only in a particular and limited way what pertains to a species in a universal and unlimited

131 Cf. *In BDH*, lect. 2, 24: "Est autem participare quasi partem capere. Et ideo quando aliquid particulariter recipit id quod ad alterum pertinet universaliter, dicitur participare illud."
132 Cf. ibid.: "Sicut homo dicitur participare animal quia non habet rationem animalis secundum totam communitatem; et eadem ratione Socrates participat hominem."
133 Cf. *DEE*, cap. 2: "Quamvis autem genus significet totam essentiam speciei, non tamen oportet ut diversarum specierum quarum est idem genus, sit una essentia, quia unitas generis ex ipsa indeterminatione vel indifferentia procedit. [...] Sed quia genus significat aliquam formam—non tamen determinate hanc vel illam—quam determinate differentia exprimit, quae non est alia quam illa quae indeterminate significabatur per genus. Et ideo dicit Commentator in XI Metaphysicae quod materia prima dicitur una per remotionem omnium formarum, sed genus dicitur unum per communitatem formae significatae. Unde patet quod per additionem differentiae remota illa indeterminatione quae erat causa unitatis generis, remanent species per essentiam diversae."; *QDSC*, a. 1, ad 2: "Ad secundum dicendum quod duplex est limitatio formae. [...] Alia vero secundum quod forma generis limitatur ad naturam speciei; et talis limitatio formae non fit per materiam, sed per formam magis determinatam, a qua sumitur differentia; differentia enim addita super genus contrahit ipsum ad speciem."

way.[134] For example, an individual human being like Angela Merkel, which participates in the species *rational animal,* has only in a particular and limited way what pertains to a species like *rational animal* in a universal and unlimited way, namely, human matter. The human matter which Angela Merkel has is designated human matter, i.e., human matter which is determined with respect to its dimensions in three-dimensional space in contrast to undesignated human matter, i.e., human matter which is undetermined with respect to its dimensions in three-dimensional space, which is common to all members of the species *rational animal* (see 6.2.2). Thus, an individual human being has in a particular and limited way human matter which pertains to the species *rational animal* in a universal and unlimited way, because the human matter which belongs to the species is undetermined with respect to its dimensions in three-dimensional space and precisely for this reason is common to all human beings.

The second way something participates in something else is how a subject participates in an accidental form and matter in a substantial form.[135] An accidental form is restricted and limited by a subject which participates in it because a subject which participates in an accidental form has only in a particular and limited way what pertains to an accidental form in a universal and unlimited way. For example, the accidental form of yellowness which inheres in a banana as its subject is not yellowness in a universal and unlimited way but only the limited and restricted yellowness of that particular banana. Yellowness as such is universal and unlimited but the yellowness *of* a particular banana is the restricted and limited yellowness of its subject. In other words, an accidental form which is participated in by a subject is restricted and limited by the subject which participates in it and with which it enters into composition because a subject individuates an accidental form.[136] Accidental forms are multiplied and diversified through different subjects in which they inhere.

A substantial form is restricted and limited by matter which participates in it because matter which participates in a substantial form has only in a particular and limited way what pertains to a substantial form in a universal and unlimited way. For example, the substantial form of a human being—a human soul—which

---

134 Cf. *QDSC*, a. 1, ad 2: "Ad secundum dicendum quod duplex est limitatio formae. Una quidem secundum quod forma speciei limitatur ad individuum, et talis limitatio formae est per materiam."

135 Cf. ibid.: "Similiter etiam subiectum participat accidens et materia formam, quia forma substantialis vel accidentalis, quae de sui ratione communis est, determinatur ad hoc vel illud subiectum."

136 Cf. *QDP*, q. 9, a. 1, ad 8: "Ad octavum dicendum, quod accidentia non individuantur nisi ex suis subiectis."

informs or configures matter and inheres in matter as its subject is not the human soul in a universal and unlimited way but the limited and restricted human soul *of* the matter it configures and inheres in, i.e., it is the human soul of a particular human being and no other. In other words, a substantial form which is participated in by matter as its subject is restricted and limited by the matter which participates in it and with which it enters into composition because matter individuates a substantial form.[137] Substantial forms are multiplied and diversified through different chunks of matter in which they inhere and with which they enter into composition.

Finally, the third way something participates in something else is how an effect whose power is not equal to the power of its cause participates in the power of its cause.[138] In such a case, the power of a cause is restricted and limited by an effect which participates in it because the effect which participates in the power of the cause has only in a particular and limited way the power which the cause has in a universal and unlimited way. Aquinas's example to illustrate this kind or mode of participation is the light of the Sun which is participated in by air. The effect of the participation of air in the light of the Sun is illuminated air, i.e., air which participates in light and enters into composition with light. But the light which is participated in by air does not have the same power to illuminate and the same degree of clarity or brightness as the light of the Sun which causes the air to be illuminated. The power to illuminate which the light of the Sun has is restricted and limited by the air in which it is received and with which it enters into composition.

Thus, in Aquinas's view, participation and metaphysical composition in general involve restriction and limitation. That which is participated in is restricted and limited by that which participates in it. Therefore, to defend Aquinas's view, there is no option to negate this premise of the objection. What about the second possibility? One could attempt to argue that the participation of a substance in *esse* is an exceptional case. The way *esse* is participated in by a substance is different from all the other ways of participation we have considered. Could it not

---

137 Cf. *DEE*, cap. 4: "Secunda differentia est quia essentiae rerum compositarum ex eo quod recipiuntur in materia designata multiplicantur secundum divisionem eius, unde contingit quod aliqua sint idem specie et diversa numero. Sed cum essentia simplicis non sit recepta in materia, non potest ibi esse talis multiplicatio; et ideo oportet ut non inveniantur in illis substantiis plura individua eiusdem speciei, sed quot sunt ibi individua tot sunt ibi species, ut Avicenna expresse dicit."
138 Cf. ibid.: "Et similiter etiam effectus dicitur participare suam causam, et praecipue quando non adaequat virtutem suae causae, puta si dicamus quod aer participat lucem solis quia non recipit eam in claritate qua est in sole."

be the case that this is a further difference, i.e., that the participation of a substance in *esse* is a special and unique case of metaphysical participation and composition in which what participates does not limit what is participated in, i.e., *esse?*

Aquinas does not adopt such a view. Rather, his view is that *esse* is restricted and limited by the essence or nature of a substance which participates in it because the essence or nature of a substance which participates in *esse* has only in a particular and limited way what pertains to *esse* in a universal and unlimited way.[139] In other words, the *esse* of a substance is restricted and limited precisely because it is participated in by the essence or nature of that substance.

Consequently, the *esse* of any substance is contracted to and limited to the *esse* of its essence or nature. For example, the *esse* of a particular human being is limited and contracted to the *esse* which members of the species *rational animal* have and the *esse* of a horse is limited and contracted to the *esse* which members of the species *horse* have.[140] Thus, in Aquinas's view, the participation of the essence or nature of a substance in *esse* is no exception to the rule that metaphysical participation and composition involve what participates restricting and limiting what is participated.

This leaves us with the third and last option for rebutting the objection that Aquinas's explanation of what it is for a substance to exist is incoherent. What must be shown is that the restriction and limitation of *esse* by the essence or nature of a substance which participates in it does not imply that the *esse* of a substance and its essence or nature are related as potentiality to actuality. It seems that *esse* is related to the essence or nature of a substance as potentiality to ac-

---

139 Cf. *QDSC*, a. 1, obj. 15: "Praeterea, omnis substantia naturae suae limitibus circumscripta, habet esse limitatum et coarctatum. Sed omnis substantia creata est naturae suae limitibus circumscripta. Ergo omnis substantia creata habet esse limitatum et coarctatum. Sed omne quod coarctatur, aliquo coarctatur. Ergo in qualibet substantia creata est aliquid coarctans, et aliquid coarctatum." It is true that this reference is taken from an objection which Aquinas considers. But Aquinas's response to this objection makes clear that he does embrace the view which is expressed in the objection. What he rejects is just the claim of the objector that one can infer from the view that *esse* is limited and contracted by the essence or nature of a substance and the view that an immaterial substance is metaphysically composed of something contracting or confining and something contracted or confined that there is a metaphysically composition of form and matter; ibid., ad 15: "Ad decimumquintum dicendum quod esse substantiae spiritualis creatae est coarctatum et limitatum non per materiam, sed per hoc quod est receptum et participatum in natura determinatae speciei, ut dictum est."

140 Cf. *QDP*, q. 1, a. 2, corp.: "Esse enim hominis terminatum est ad hominis speciem, quia est receptum in natura speciei humanae; et simile est de esse equi, vel cuiuslibet creaturae."

tuality because *esse* which is participated in is related to the essence or nature of a substance which participates in it as a determinable to a determinant. The essences or natures of different species determine and individuate *esse* to the *esse* of a human being, the *esse* of a horse, or the *esse* of a stone etc.[141] The question is how Aquinas's view reconciles the claim that *esse* is determined and limited to the *esse* of the essence or nature of a substance with the claim that *esse* determines the essence or nature of a substance—and consequently the substance itself taken as what it is essentially—with respect to its actuality.

In a nutshell, Aquinas's solution to this puzzle runs as follows: He affirms that *esse* is determined and limited to the *esse* of the essence or nature of a substance which participates in it. Yet he negates that it follows from this that *esse* which is participated in is related to the essence or form of a substance which participates in it as potentiality to actuality. In other words, he negates that the essence or nature of a substance determines the *esse* in which it participates as actuality determines potentiality. Rather, the essence or nature of a substance determines the *esse* in which it participates as potentiality determines actuality.[142]

Therefore, Aquinas's view is not incoherent or inconsistent because the claim that the essence or nature of a substance determines the *esse* in which it participates as potentiality determines actuality does not contradict the claim that *esse* determines the essence or nature of a substance which participates in it—and consequently the substance itself taken as what it is essentially—with respect to its actuality. In other words, there is no incoherence or inconsistency because the claim that *esse* determines the essence or nature of a substance which participates in it as actuality determines potentiality does not conflict with the claim that the essence or nature of a substance determines the *esse* it participates in as potentiality determines actuality. What must be distinguished are two kinds or modes of determination which take place in the participation of a substance in *esse* which do not contradict but rather complement each other.

This brief sketch of Aquinas's solution requires a bit of unfolding. Let us begin with the question of why the essence or nature of a substance does not determine *esse* as actuality determines potentiality. The essence or nature of a

141 Cf. *SCG*, II, cap. 52: "Esse autem, inquantum est esse, non potest esse diversum: potest autem diversificari per aliquid quod est praeter esse; sicut esse lapidis est aliud ab esse hominis."

142 Cf. *QDP*, q. 7, a. 2, ad 9: "Nec intelligendum est, quod ei quod dico esse, aliquid addatur quod sit eo formalius, ipsum determinans, sicut actus potentiam. [...] Unde non sic determinatur esse per aliud sicut potentia per actum, sed magis sicut actus per potentiam."

substance does not determine *esse* as actuality determines potentiality because *esse* is not participated in by the essence or nature of a substance as a genus is participated in by a species for the reasons we have seen above in detail (see 6.2.2).[143] The view that the essence, nature, or form of a substance which participates in *esse* determines *esse* which is participated in as actuality determines potentiality assumes that *esse* is like prime matter which is completely undetermined with respect to form and which gets determined by any form in which it participates.[144] But according to Aquinas, such a view makes no sense because there is nothing outside of *esse* which could be added to *esse* such that it determines *esse* with respect to its actuality.[145]

The following analogy might help to see the objector's mistake: The accidental form of yellowness which is participated in by a particular banana is not determined by the particular banana which participates in it with respect to its yellowness. The banana which participates in yellowness does not add any yellowness to yellowness, i.e., it does not make yellowness more yellow. In an analogous way, *esse* which is participated in by a particular form is not determined by the particular form which participates in it with respect to its *esse* or actuality. The form which participates in *esse* does not add any *esse* or actuality to *esse*, i.e., it does not make *esse* more *esse* or actual.

From Aquinas's point of view, the objection starts from a false premise, namely, that *esse* is related to form as potentiality to actuality. The opposite is true: *Esse* is related to form as actuality to potentiality.[146] *Esse* is the highest per-

---

143 Cf. *QDP*, q. 7, a. 2, ad 9: "Nec intelligendum est, quod ei quod dico esse, aliquid addatur quod sit eo formalius, ipsum determinans, sicut actus potentiam: esse enim quod huiusmodi est, est aliud secundum essentiam ab eo cui additur determinandum. Nihil autem potest addi ad esse quod sit extraneum ab ipso, cum ab eo nihil sit extraneum nisi non-ens, quod non potest esse nec forma nec materia. Unde non sic determinatur esse per aliud sicut potentia per actum, sed magis sicut actus per potentiam."

144 Cf. *QDSC*, a. 1, corp.: "Manifestum est enim quod cum potentia et actus dividant ens, et cum quodlibet genus per actum et potentiam dividatur; id communiter materia prima nominatur quod est in genere substantiae, ut potentia quaedam intellecta praeter omnem speciem et formam, et etiam praeter privationem; quae tamen est susceptiva et formarum et privationum."; *QDP*, q. 7, a. 2, obj. 9: "Sed esse est imperfectissimum, sicut prima materia: sicut enim materia prima determinatur per omnes formas, ita esse, cum sit imperfectissimum, determinari habet per omnia propria praedicamenta."

145 Cf. *QDP*, q. 7, a. 2, ad 9: "Nihil autem potest addi ad esse quod sit extraneum ab ipso, cum ab eo nihil sit extraneum nisi non-ens, quod non potest esse nec forma nec materia."

146 Cf. *ST*, I, q. 4, a. 1, ad. 3: "Ad tertium dicendum quod ipsum esse est perfectissimum omnium, comparatur enim ad omnia ut actus. Nihil enim habet actualitatem, nisi inquantum est. Unde ipsum esse est actualitas omnium rerum, et etiam ipsarum formarum. Unde non comparatur ad alia sicut recipiens ad receptum, sed magis sicut receptum ad recipiens. Cum enim

fection of all; it is the actuality of all actualities and the perfection of all perfections, i.e., *esse* is related to every formal perfection of a substance as actuality to potentiality.[147] In other words, every formal perfection of a substance is a formal perfection which can give *esse* to a substance only because the formal perfection itself has *esse* and is actual.[148]

So far, I have explained why it is ruled out by Aquinas that *esse* is determined by the essence, nature, or form of a substance which participates in it as potentiality is determined by actuality. What remains to be explained is how *esse* is determined by the essence or nature of a substance which participates in it as actuality is determined by potentiality. Aquinas explains this kind or mode of determination by referring to the way a substantial form is determined and limited by the matter which participates in it.[149]

As we have seen above, a substantial form is determined and limited by matter which participates in it because matter which participates in a substantial form has only in a particular and limited way what pertains to a substantial form in a universal and unlimited way. For example, the substantial form of a human being—a human soul—which configures matter and inheres in matter as its subject is not the human soul in a universal and unlimited way but the lim-

---

dico esse hominis, vel equi, vel cuiuscumque alterius, ipsum esse consideratur ut formale et receptum, non autem ut illud cui competit esse."

147 Cf. *QDP*, q. 7, a. 2, ad 9: "Ad nonum dicendum, quod hoc quod dico esse est inter omnia perfectissimum: quod ex hoc patet quia actus est semper perfectio potentia. Quaelibet autem forma signata non intelligitur in actu nisi per hoc quod esse ponitur. Nam humanitas vel igneitas potest considerari ut in potentia materiae existens, vel ut in virtute agentis, aut etiam ut in intellectu: sed hoc quod habet esse, efficitur actu existens. Unde patet quod hoc quod dico esse est actualitas omnium actuum, et propter hoc est perfectio omnium perfectionum."; *QDSC*, a. 1, corp.: "Remoto igitur fundamento materiae, si remaneat aliqua forma determinatae naturae per se subsistens, non in materia, adhuc comparabitur ad suum esse ut potentia ad actum: non dico autem ut potentiam separabilem ab actu, sed quam semper suus actus comitetur. Et hoc modo natura spiritualis substantiae, quae non est composita ex materia et forma, est ut potentia respectu sui esse."

148 Cf. *ST*, I, q. 4, a. 2, corp.: "Omnium autem perfectiones pertinent ad perfectionem essendi, secundum hoc enim aliqua perfecta sunt, quod aliquo modo esse habent."; *ST*, I, q. 4, a. 1, ad. 1: "Sed quia in his quae fiunt, tunc dicitur esse aliquid perfectum, cum de potentia educitur in actum, transumitur hoc nomen perfectum ad significandum omne illud cui non deest esse in actu, sive hoc habeat per modum factionis, sive non."; ibid., corp.: "Secundum hoc enim dicitur aliquid esse perfectum, secundum quod est actu, nam perfectum dicitur, cui nihil deest secundum modum suae perfectionis."

149 Cf. *QDP*, q. 7, a. 2, ad 9: "Unde non sic determinatur esse per aliud sicut potentia per actum, sed magis sicut actus per potentiam. Nam et in definitione formarum ponuntur propriae materiae loco differentiae, sicut cum dicitur quod anima est actus corporis physici organici."

ited and restricted human soul *of* the matter it configures and inheres in, i.e., it is the human soul of a particular human being and no other. Matter which participates in a substantial form individuates the substantial form it participates in.[150] Substantial forms are multiplied and diversified through different chunks of matter in which they inhere and with which they enter into composition.

According to Aquinas, *esse* is not a substantial form and an essence or form is not matter but *esse* is determined by the essence or nature of a substance which participates in it and which enters into composition with it in an analogous way. In analogy to matter which participates in a substantial form and thereby determines, limits, and individuates the substantial form it participates in, the essence or form of a substance which participates in *esse* determines, limits, and individuates the *esse* it participates in to the particular *esse* of this or that essence or nature, e.g., the *esse* of a human being, the *esse* of a horse, the *esse* of a stone etc.[151] Matter does not determine the substantial form it participates in and with which it enters into composition with respect to its actuality by determining it in this way. In a similar way, the essence or nature of a substance does not determine the *esse* it participates in and with which it enters into composition with respect to its actuality by determining it in this way.

To sum up: *Esse* is participated in by the essence or nature of a substance and determines the essence or nature of a substance—and consequently the substance itself taken as what it is essentially—with respect to its actuality. The essence or nature of a substance which participates in *esse* determines *esse* to the *esse* of the essence or nature of the substance. Thus, metaphysical participation and composition involve limitation and restriction of that which is participated in through that which participates. The participation of the essence or form of a substance in *esse* is no exception. Nevertheless, these two claims do not contradict each other because we are dealing with two kinds or modes of determination. *Esse* determines the essence or nature of a substance as actuality deter-

---

150 Cf. *DEE*, cap. 4: "Secunda differentia est quia essentiae rerum compositarum ex eo quod recipiuntur in materia designata multiplicantur secundum divisionem eius, unde contingit quod aliqua sint idem specie et diversa numero. Sed cum essentia simplicis non sit recepta in materia, non potest ibi esse talis multiplicatio; et ideo oportet ut non inveniantur in illis substantiis plura individua eiusdem speciei, sed quot sunt ibi individua tot sunt ibi species, ut Avicenna expresse dicit."; *QDSC*, a. 1, ad 2: "Ad secundum dicendum quod duplex est limitatio formae. Una quidem secundum quod forma speciei limitatur ad individuum, et talis limitatio formae est per materiam."

151 Cf. *QDP*, q. 7, a. 2, ad 9: "Et per hunc modum, hoc esse ab illo esse distinguitur, in quantum est talis vel talis naturae."; *QDP*, q. 1, a. 2, corp.: "Esse enim hominis terminatum est ad hominis speciem, quia est receptum in natura speciei humanae; et simile est de esse equi, vel cuiuslibet creaturae."

mines potentiality and the essence or nature of a substance determines *esse* as potentiality determines actuality. Consequently, Aquinas's view is perfectly coherent and consistent.

7 Why Aquinas's View Can Solve the Problems

In the previous chapter, I have reconstructed and defended Aquinas's view on what it is for a substance to exist. Aquinas's final answer to the nature question is that the existence of a substance consists in its metaphysical composition of a determinable and a determinant, namely, its essence or nature and its *esse*. In this final chapter, I will explain in detail why Aquinas's view can solve the problems which the views discussed in the first part of my book cannot.

The views I have discussed in the first part can be divided into two camps. The non-reductive and the reductive-object view have in common the idea that the existence of an object consists in that *object*'s having a property. The remaining three views reject this way of explaining what it is for an object to exist. They have in common the idea that an object's existence consists in *something else*'s having a property. I will begin by explaining why Aquinas's view can solve the problems of the three views which belong to the latter camp.

7.1 Why Aquinas's View Can Solve the Problems of the Property View and the Domain Views

7.1.1 Solution to a Vicious Circularity

The property view, the unrestricted-domain view, and the restricted-domain view have one thing in common. According to these three views, an object's existence does not consist in a property of that *object* (e. g., the property of existing, or the property of being numerically identical with itself) but in a property of *something else* (e. g., an unrestricted or restricted domain of objects' property of containing the object, or a property's property of being instantiated by that object). Thus, their different explanations for what it is for an object to exist have a common form: The existence of an object X consists in something else Y's having a property F.

The three views only differ with respect to what 'Y' and 'F' stand for. 'Y' stands either for a property or a domain of objects (which in turn is either unrestricted or restricted) and 'F' stands either for a property's property of being instantiated or the property of a domain of objects (which is either unrestricted or restricted) of containing X and being [X]ish. The views are committed to the idea that F is a property of something other than X, namely Y, due to their rejection of the claim that the existence of X consists in X's having a property. Since Y's having F is supposed to explain the existence of X, the former cannot presuppose

https://doi.org/10.1515/9783110979879-010

the latter. An explanation for the existence of X which presupposes the existence of X is viciously circular and explains nothing.

The explanations of what it is for an object X to exist, which the three views have to offer, have precisely this problem. Let us begin with the property view. According to this view, 'Y' stands for a property and 'F' stands for Y's property of being instantiated by the object X whose existence is supposed to be explained. Thus, the existence of an object consists in a property's having the property of being instantiated. A property's having the property of being instantiated is supposed to explain the existence of an object X. In other words, the existence of X should be the result of a property's having the property of being instantiated if the property's having the property of being instantiated is supposed to explain the existence of X.

The problem is that a property's having the property of being instantiated does not explain but rather presupposes the existence of the object whose existence is supposed to be explained. A property Y can only have the property of being instantiated if there exists at least one instance of Y, i.e., one object by which Y is instantiated. Since Y's having the property of being instantiated is supposed to explain the existence of an object X, this instance of Y, i.e., the object by which Y is instantiated, must be X. Yet as I have shown above with the example of the mistakenly posited inter-Mercurial planet called 'Vulcan' (see 3.2.1), X must exist to be an instance of Y. Thus, Y's having the property of being instantiated by X does not explain what the existence of X consists in but rather presupposes it.

Even worse, the explanatory order is the reverse. The existence of X is not the result of Y's having the property of being instantiated by X but Y's having the property of being instantiated by X is the result of the existence of X. In other words, X does not exist in virtue of a property Y's having the property of being instantiated by X, but Y has the property of being instantiated by X in virtue of X's existence. Thus, instantiation is explained in terms of existence and not the other way around.

The unrestricted-domain view and the restricted-domain view fare no better. In their case, 'Y' stands for an unrestricted domain of objects—a UDO—or a restricted domain of objects—a REDO—and 'F' stands for the property of a UDO or a REDO of containing an object X and being [X]ish. The answers to the nature question which these two views have to offer are viciously circular for a similar reason. A domain of objects Y's having the property of containing an object X and being [X]ish does not explain but presupposes the existence of X. Y can only have the property of containing X and being [X]ish if X exists. If X does not exist, Y cannot have the property of containing X and being [X]ish. Again, the explanatory order is the reverse: It is not Y's having the property of containing X and being [X]ish which explains what it is for X to exist. Rather, the existence of X must be presupposed to make sense of the idea that Y has the property of containing X and being [X]ish.

Against this background, it is easy to see how Aquinas's view can solve the problem of a vicious circularity in the form in which it arises for the three views. The problem of a vicious circularity does not arise for Aquinas's view in this form because it rejects two premises which give rise to the problem. First, Aquinas's view rejects the claim that the existence of an object X consists in *something else Y's* having a property F. According to Aquinas's view, the existence of an object X —a substance in his terminology—consists in X's having existence or actuality—*esse* in his terminology. That which accounts for the existence of X, namely, the *esse* of a substance, belongs to X and not to Y. There is no need to introduce a further entity Y to explain what it is for X to exist.[1] The existence of X is the result of X's metaphysical composition of a determinable and a determinant, namely, its essence or nature and its *esse*.

Second, Aquinas's view rejects the premise that existence or *esse* can be identified with and reduced to a *property* of Y. According to Aquinas, *esse* is neither a property of an object X nor of something else Y. Rather, *esse* is primitive, i.e., it cannot be identified with and reduced to anything else. Thus, the problem of a vicious circularity is solved by Aquinas's view by rejecting the premises which give rise to the problem.

7.1.2 Solution to a Vicious Infinite Regress

The second problem that the property view, the unrestricted-domain view, and the restricted-domain view have in common is the problem that their way of explaining what it is for an object to exist triggers a vicious infinite regress. Again, this problem is generated by the common form of their different explanations for what it is for an object X to exist: The existence of an object X consists in something else Y's having a property F. It does not matter whether 'Y' stands for a property or a domain of objects (either unrestricted or restricted) and whether 'F' stands for the property of a property of being instantiated or the property

1 In Aquinas's view, a further 'entity' or *ens*, namely, subsistent *esse* itself or God, is required to explain in virtue of what a substance has the *esse* it has. In other words, Aquinas's answer to the strong ground question requires the existence of a further entity or *ens* Y, but not Aquinas's answer to the weak ground question or nature question. Now, it is important to note that the term 'entity' or '*ens*' is used in an analogous sense here. In Aquinas's view, God is not an entity or *ens* in the same sense as any of the entities we are familiar with.

of a domain of objects (either unrestricted or restricted) of containing X and being [X]ish. Any view which tries to explain what it is for an object X to exist in this way triggers a vicious infinite regress because it must presuppose that Y exists.

Otherwise, the explanation of what it is for X to exist does not work. Y's existence must be presupposed because something which does not exist itself cannot have the property of being instantiated or the property of containing X and being [X]ish. *Having* the property of being instantiated and *having* the property of containing X and being [X]ish requires that Y as the subject or bearer of that property exists. It would be absurd to claim that the existence of X consists in something Y's having a property if Y does not exist. Thus, the explanation of what it is for X to exist must presuppose the existence of Y.

But this means that the explanation of the three views of what it is for an object X to exist is unsatisfying unless an explanation is given for what it is for Y to exist, since Y plays an important part in the views' explanation of what it is for X to exist. Now, according to the property view, the unrestricted-domain view, and the restricted-domain view, that which exists does not exist in virtue of having a property but rather in virtue of *something else Y's* having a property. Consequently, any such view must explain what it is for Y to exist in the following way: The existence of Y consists in something else Z's having a property. For example, if Y is a property, the explanation of the view of what it is for Y to exist is that another property Z has the property of being instantiated. And if Y is a domain of objects, the explanation of the view of what it is for Y to exist is that another domain of objects Z has the property of containing Y and being [Y]ish.

However, the problem just reiterates itself because this way to explain what it is for Y to exist presupposes the existence of Z and so on *ad infinitum*. This infinite regress is vicious because it becomes impossible for the view to formulate a satisfactory explanation of what it is for X to exist. The existence of Y plays an important part in the explanation of what it is for X to exist which makes it necessary to explain what it is for Y to exist to have a satisfying explanation of what it is for X to exist. Now, the existence of Z plays an important part in the explanation of what it is for Y to exist which makes it necessary to explain what it is for Z to exist to provide a satisfying explanation of what it is for Y to exist, whose existence plays an important part in the explanation of what it is for X to exist and so on *ad infinitum*.

Aquinas's view can solve the problem of a vicious infinite regress in the same way as it solves the problem of the vicious circularity. Aquinas's view negates the premise which triggers the infinite regress. What is negated is the claim that the existence of an object X consists *in something else Y's* having a property.

In Aquinas's view, for an object X to exist is to have *esse*. *Esse* belongs to X and not to Y. Consequently, Aquinas's answer to the nature question works without the introduction of a further entity Y the existence of which must be presupposed. Put otherwise, what explains the existence of X is that X is metaphysically composed of a determinable—its essence or nature—and a determinant—its *esse* —and these metaphysical constituents or components do not exist prior to the whole they compose, i.e., the determinate. Furthermore, they exist as metaphysical parts of the substance, i.e., they exist only in and with the substance as that by which (*quo est*) the substance exists. Thus, no infinite regress is triggered.[2]

7.1.3 Solution to Ontological Inflationism

A third problem which is common to the property view and the restricted-domain view is the problem that their explanations of what it is for an object X to exist work only with a significantly inflated ontology. The problem is generated through the combination of two claims. First, the views maintain that the existence of an object X consists in something else Y's having a property F. Second, the views assume that Y is something restricted or limited whose having F does not account for the existence of every object that exists.

Let us begin with the property view according to which the existence of an object X consists in a property Y's having the property of being instantiated by X. First, this explanation of what it is for X to exist must presuppose that Y exists itself. It would be absurd to claim that the existence of an object X consists in a property Y's having the property of being instantiated if Y does not exist. Y can be the subject or bearer of the property, which is supposed to explain the existence of X, only if Y exists. Thus, to explain what it is for X to exist, the view needs to introduce Y into its ontology.

---

2 It is important to note that the fact that these metaphysical parts exist does not imply that they are *additional* entities, i.e., entities (e.g., Y, and Z) which exist *in addition to* X, and whose existence is presupposed by the explanation of the existence of X which is given. They exist only in and with X as that by which (*quo est*) X exists. In other words, they are not entities in the sense of that which is (*id quod est*) but only entities in the qualified sense of that by which (*quo est*) that which is is. It is important not to reify metaphysical principles like essence or *esse* because this would trigger an infinite regress. The problem with the property-view, the unrestricted-domain view, and the restricted-domain view is that they require and presuppose the existence of an additional entity Y which is an entity in the sense of that which is (*id quod est*) to explain what it is for X to exist. Thanks to Timothy Pawl for raising this worry. I am also thankful to Jeremy Skrzypek for making me aware of the need to expand my thoughts a bit at this place to dispel the possible worry that Aquinas's view itself is vulnerable to the objection.

Second, the property Y's having the property of being instantiated explains only what it is for X to exist if Y can be instantiated only by X. For example, the explanation of what it is for Angela Merkel to exist cannot be that the property of being a human being or the property of humanity has the property of being instantiated because humanity can have the property of being instantiated without it being the case that Angela Merkel exists. If Angela Merkel dies, the property of being a human being or the property of humanity does not lose the property of being instantiated because there are still many other human beings by which this property is instantiated. Consequently, the relevant property Y, whose having the property of being instantiated is supposed to explain what it is for X to exist, must be restricted or limited in such a way that Y can be instantiated only by X. For example, to explain what it is for Angela Merkel to exist, the property view must claim that there exists a property which individuates Angela Merkel, i.e., a property of which Angela Merkel is the only possible instance. In the case of Angela Merkel, it could be argued that the relevant property is the property of being a woman named 'Angela Merkel' who was born on the 17th of July 1954 in Hamburg and who is the 8th chancellor of Germany after the Second World War. The existence of Angela Merkel consists in this property's having the property of being instantiated. It follows that the property view must introduce into its ontology at least as many properties as those required to individuate all the objects that exist to explain what it is for these objects to exist.[3]

The restricted-domain view must inflate its ontology for similar reasons. First, to explain what it is for an object X to exist, the view must presuppose the existence of a restricted domain of objects Y. It would be absurd to claim that the existence of X consists in a non-existing restricted domain of objects Y's having the property of containing X and being [X]ish. Something which does not exist cannot have the property of containing X and being [X]ish. Thus, the view must admit Y into its ontology to be able to explain what it is for X to exist.

Second, since the restricted-domain view is committed to the claim that the domain of objects Y, whose having the property of containing X and being [X]ish is supposed to explain what it is for X to exist, is *restricted* and *limited*, the view

---

3 Again, it could be argued that the property of Angela Merkel is a composite property, i.e., a bundle of properties each of which are instantiated by other things. So, in that case, we only need to suggest that the property of Angela Merkel is a unique combination of already instantiated properties. In this way, the property view does not need to introduce additional properties. This might be true. But again, it is at best a trade-off because thereby the property view loses the ability to explain rather easily what individuates an object like Angela Merkel. Thanks to Jeremy Skrzypek for raising this worry.

has to introduce not only one but a *plurality* of restricted domains of objects into its ontology. Thus, this approach must inflate its ontology with a significant number of restricted or limited domains of objects to explain what it is for the plurality of objects that exist to exist.[4]

The problem of ontological inflationism worsens considerably in combination with the problem of a vicious infinite regress. If an infinite regress is triggered, an explanation of what it is for a *single* object to exist requires admitting an *infinite* number of existing restricted or limited domains of objects into one's ontology.[5] Nevertheless, the problem of ontological inflationism is a problem of its own which can be considered independently from the problem of a vicious infinite regress.[6] The need to inflate one's ontology with further entities—properties or domains of objects—does not vitiate the aim of the property view and the restricted-domain view to formulate a satisfying explanation for what it is for an object to exist as a vicious infinite regress does.

An explanation which requires a considerably inflated ontology to explain what it is for an object to exist is not damaged beyond repair as is the case with an explanation which triggers a vicious infinite regress. Nevertheless, a considerably inflated ontology is clearly a metaphysical price or cost and makes an answer to the nature question less attractive. A view which admits, besides objects, a considerable number of properties or restricted domains of objects as entities into its ontology which exist in the same sense as objects is less simple, parsimonious, and elegant than a view which admits only objects as truly and properly existing entities into its ontology.

Aquinas's view solves the problem of ontological inflationism, which arises for the property view and the restricted-domain view, by rejecting the two claims which give rise to the problem. In Aquinas's view, the existence of an object X—a

---

4 It could be argued that the explanation of the restricted-domain view for what it is for the objects that exist to exist might work with a less inflated ontology than the explanation which is given by the property view, because it can be the case that a restricted domain of objects Y has the property of containing several objects. For example, by introducing the restricted domain of objects Y into its ontology, the restricted-domain view can explain what it is for objects A, B, and C to exist if Y has the property of containing A, B, and C and is [A]ish, [B]ish, and [C]ish. This might be true. Nevertheless, it is still the case that the ontology of this view is inflated. It contains not only the *objects* which exist but also all the *restricted domains of objects* which are required to explain what it is for the objects that exist to exist.

5 Gabriel is willing to accept this consequence of his answer to the nature question, cf. Gabriel, *Fields of Sense*, 159 – 160, 196, 242; "The Meaning of 'Existence' and the Contingency of Sense," 127 – 128.

6 As I have argued above (see 5.2.1 and 5.2.2), for this reason Gabriel is wrong in believing that he has solved the problem of a vicious infinite regress by embracing a heavily inflated ontology.

substance in his terminology—does not consist in something else Y's having a property F. Rather, the existence of a substance consists in the substance's having existence or actuality (*esse* in his terminology). *Esse,* which accounts for the existence or actuality of this substance, belongs to the substance and not to something else. The existence of a substance results from its metaphysical composition of a determinable and a determinant, namely, its essence or nature and its *esse.* Consequently, there is no need for Aquinas's view to introduce further entities like properties or restricted domains of objects into its ontology which exist in the same sense as substances.

Paradoxically, it is the attempt of the property view and the restricted-domain view to identify and *reduce* the existence of an object X with and to a property something else Y has which requires them to inflate their ontologies with *additional* entities like properties or restricted domains of objects which exist in their own right. Aquinas's view can evade such an inflation of its ontology by acknowledging that *esse* is primitive and belongs to substances as that which determines them with respect to their actuality. In other words, it is Aquinas's unwillingness to reduce and identify *esse* with anything else which allows him to embrace a parsimonious ontology according to which the only objects which exist in a proper and unqualified sense are substances which exist *per se* and *in se.*[7]

7 It could be objected that Aquinas's view is not more parsimonious than the property view and the restricted-domain view because Aquinas's view also introduces several new kinds of entities (e.g., prime matter, substantial form, essence, *esse* etc.) to answer the nature question. The problem with this objection is that it does not take into account that in Aquinas's view these kinds of entities do not exist *in their own right.* As I have pointed out above (see. 6.1), according to Aquinas such kinds of entities are only entities in the derived and qualified sense of being metaphysical parts or constituents of substances. They are entities only in the sense that they exist in and with substances as that by which (*quo est*) substances have substantial and accidental existence. Thus, Aquinas's view is more parsimonious than the property view and the restricted-domain view in the sense that it does not need to introduce several new kinds of entities which exist in their own right to explain what it is for substances to exist. According to Aquinas, it is important to note that '*ens*' is not a univocal but an analogical term which is attributed in a different sense to a substance and a metaphysical part or constituent of a substance, cf. *QDV,* q. 21, a. 4, ad. 4: "Aliquid enim dicitur ens, quia in se subsistit; aliquid, quia est principium subsistendi, ut forma; aliquid, quia est dispositio subsistentis, ut qualitas; aliquid, quia est privatio dispositionis subsistentis, ut caecitas. Et ideo cum dicimus: essentia est ens; si procedatur sic: ergo est aliquo ens, vel se vel alio; processus non sequitur, quia non dicebatur hoc modo esse ens, sicut aliquid subsistens in esse suo est ens, sed sicut quo aliquid est. Unde non oportet quaerere quomodo ipsa essentia aliquo sit sed quomodo aliquid alterum sit per essentiam." Nevertheless, it is a fair point that Aquinas's view is not more parsimonious in the sense that it requires the existence of metaphysical parts or constituents of substances—which are entities only

in a qualified and derived sense—to explain what it is for substances to exist. It can be argued that this is an attractive trade-off because with the help of the introduction of metaphysical parts or constituents Aquinas's view is able to provide a more satisfactory answer to the nature question than the considered rival views which reject the existence of such entities. I am thankful to Jeremy Skrzypek for raising this objection. Timothy Pawl expressed the worry that this rebuttal is still not satisfying because things like prime matter, substantial form, essence, *esse* etc. must be *real*, they must exist in some way or other, and they must be really *distinct* from each other to play the roles that Aquinas needs them to in his theory. For instance, matter must be really *distinct* from form and both must *really* be there as metaphysical constituents in order to explain change across time. So, according to Pawl, they should count as ontological additions. However, if they are ontological additions, it seems like an unfair counting strategy to say that they don't *really* count whereas, for example, REDOs do count. I am thankful to Pawl for pushing me here because it helps me to clarify an important point. It is true that prime matter, substantial form etc. are *real*, i.e., that they exist in some way or other. Furthermore, it is true that they are really distinct from another, e.g., prime matter is really distinct from substantial form, and essence is really distinct from *esse*. However, Pawl's conclusion follows only if one accepts an additional premise, namely, that every real distinction implies separability of that which is distinguished. In other words, if it were true that whenever two things are really distinct, they are separable, it would follow that metaphysical parts like prime matter, substantial form, essence, and *esse* etc. are ontological additions in Pawl's sense. But not every real distinction does imply the separability of that which is distinguished. For example, a circle has the properties having a radius and having a circumference. These properties are really distinct properties of the circle. Nevertheless, they are not separable. You cannot have a circle with a radius without having a circumference. Thus, one cannot infer without further argument from the real distinction of these metaphysical parts that they are separable and therefore ontological additions. Maybe the following analogy is helpful to illustrate the point. In Aquinas's view, accidental forms are metaphysical parts of substances. So, they are *real* insofar as they are that by which a substance has accidental *esse*, i.e., is modified in a certain accidental way. Furthermore, an accidental form is *really distinct* from the substance in which it inheres. However, although an accidental form is really distinct from the substance in which it inheres, it is not separable from that substance. In Aquinas's view, there are no separate forms like redness, roundness, or humanity. Thus, according to Aquinas's metaphysical framework, accidental forms are no ontological additions to substances in Pawl's sense. In an analogical way, the same holds for prime matter, substantial form, essence, and *esse*. As metaphysical parts of a material substance they are *real* insofar as they are that by which a substance has substantial *esse*. And they are *really distinct* from another. Prime matter is really distinct from substantial form, and essence is really distinct from *esse*. However, they are not separable from each other because they relate as potentiality to actuality and potentiality and actuality are not separable. Furthermore, the relevant metaphysical parts are not separable from the whole in which they inhere—I leave aside the possible exception of the disembodied human soul—because they have existence only in and with the whole as that by which the whole has actuality as what it is. It is important to note that the case with, for example, REDOs is different. They count as ontological additions in Pawl's sense because a proponent of the restricted-domain view must claim that both relata of the relation are real, really distinct from each other, and that REDOs exist in separation from the objects whose existence they are supposed to explain. Thus, there is nothing unfair in counting entities like REDOs as ontological additions but not the metaphysical parts of substances because only the former view introduces

7.1.4 Solution to the Problem of Co-Existence

The fourth problem is a problem which is common to the property view, the unrestricted-domain view, and the restricted-domain view. These three views have the common problem that an object X cannot co-exist with the entity Y whose having a property F is supposed to explain what it is for X to exist. To solve this problem, the three views must introduce different ontological realms into their ontologies. Again, the problem of co-existence is generated by the common form of their different explanations of what it is for an object X to exist: The existence of an object X consists in something else Y's having a property F.[8]

As we have already seen in the previous sections, if the existence of an object X consists in something else Y's having a property F, Y must exist. Thus, X exists, and Y exists. But *where* does Y exist? Y cannot simply co-exist with X, i.e., Y cannot exist *alongside* X in the realm of objects, because this would amount to the view that the existence of an object X consists in another *object* Y's having a property F. The latter claim must be rejected by the three views because it conflicts with their claim that existence is not a property of objects.

Furthermore, if Y were another object which simply co-exists alongside X, Y could not have the property of being instantiated which is supposed to explain what it is for X to exist because objects are usually distinguished from properties on the grounds that the former cannot be instantiated.[9]

In the case of the unrestricted-domain view and the restricted-domain view, it could be argued that Y itself is a kind of object. But if a domain of objects Y is itself an object, it would follow that the existence of X consists in another object Y's having the property of containing X. In other words, the existence of X would consist in being a mereological part of object Y. Consequently, someone who entertains such a view would have to give up the idea that objects like human beings, dogs, or cats exist in their own right as complete or whole entities, i.e., entities which exist *per se* and *in se*. Instead, such a person would be forced to

---

further entities which exist in separation from the entities whose existence is supposed to be explained, i.e., REDOs.

8 In the first part of my book, I have shown only that the property view has this problem (see 3.2.4). Nevertheless, the unrestricted-domain view and the restricted-domain view also face this problem due to their way of explaining what it is for an object X to exist, as will become evident in what follows.

9 Cf. McGinn, *Logical Properties*, 15–16; Nelson, "Existence."

embrace the absurd view that any object X that seems to exist *per se* and *in se*, in truth exists only as an integral part of another object Y.[10]

To evade these consequences, views which argue that the existence of an object X consists in something else Y's having a property F must reject the idea that X co-exists with Y. Objects do not co-exist with the properties or the domain(s) of objects whose having the property of being instantiated or of containing them is supposed to explain what it is for the objects to exist. Accordingly, the three views must claim that X and Y exist but that they do not co-exist. They do not exist alongside each other. Rather, they exist in different ontological realms. In this way, the problem of the co-existence of X and Y is solved.

The price of this solution is that the unity of mind-independent reality must be given up. Instead, mind-independent reality is at least divided up into an ontological realm R_1 in which the objects exist and an ontological realm R_2 in which the properties or domains of objects exist whose having a property F is supposed to explain what it is for the objects in R_1 to exist. Thus, the strategy which the three views employ to explain what the existence of an object consists in only works with an ontology which divides up mind-independent reality at least into an ontological realm R_1 of objects and an ontological realm R_2 of properties or domains of objects.

It is important to note that I do not claim that this solution to the problem of co-existence makes the three views absurd. As with the problem of ontological inflationism, my aim is more modest. All I want to show is that this way to explain what it is for an object to exist comes at a considerable metaphysical cost or price because it only works with an ontology which divides up mind-independent reality into at least two different ontological realms. This division of mind-independent reality into two or more different ontological realms confronts any view which adopts such an explanatory strategy with a couple of well-known problems. For example, it is notoriously difficult to understand what kind of relation between an object as a denizen of one ontological realm and a property or domain of objects as a denizen of a different ontological realm could explain that an object has a certain characteristic or property.[11]

Furthermore, if objects which pertain to the ontological realm R_1 exist *in virtue of* standing in relation to a property or a domain of objects which belongs to another ontological realm R_2, it becomes difficult to avoid the consequence that

10 Recent proponents of the restricted-domain view, like Markus Gabriel, have explained at length why such a mereological interpretation of the view does not work and why domains of objects do not simply co-exist alongside objects, cf. Gabriel, *Fields of Sense*, 253–262.
11 For a detailed exposition of this problem, see, for example, Koslicki, *Form, Matter, Substance*, 15–17; Lowe, "A Neo-Aristotelian Substance Ontology," 234–235.

inhabitants of R_1 only have a diminished ontological status because their existence is dependent on the existence of the inhabitants of R_2. If what it is for objects to exist is explained in this way, there is a danger that we end up with a view according to which objects are a kind of second-class entities (see 5.2.4 and 7.1.5).

Aquinas's view does not require dividing up mind-independent reality into different ontological realms to solve the problem of the co-existence of an object X and the entity Y, whose having a property F is supposed to account for the existence of X, because Aquinas's way to explain what it is for an object X to exist does not generate this problem. Again, the problem does not arise for Aquinas's answer to the nature question because it negates the premise which generates the problem, namely, the idea that the existence of a substance X consists in something else Y's having a property F. Aquinas's view does not require introducing a further entity Y which exists itself to explain what it is for a substance X to exist. For example, the existence of Angela Merkel does not consist in something else's having a property. Rather, the existence of Angela Merkel consists in Angela Merkel's having *esse*. She exists in virtue of having *esse* and not in virtue of something else Y's having a property.

7.1.5 Solution to the Problem of Constitution

The fifth and final problem which Aquinas's view can solve is specific to the restricted-domain view. The problem is that a restricted domain of objects Y's having the property of containing an object X and being [X]ish is not constitutive of the existence of X. Once more, the problem is generated by the general form of explanation of this view of what it is for an object X to exist: The existence of an object X consists in something else Y's having a property F.

That an explanation of what it is for an object X to exist with this form does not work becomes evident if we substitute 'X' for an ordinary substance, i.e., an object which exists *per se* and *in se*. For example, let 'X' stand for Joe Biden. Whatever a restricted domain of objects or context 'Y' stands for, no restricted domain of objects or context is such that its having the property of containing Joe Biden and being [Joe Biden]ish is constitutive of Biden's existence. Any Y's having the property of containing Joe Biden and being [Joe Biden]ish is not constitutive of Biden's existence because Y's having the property of containing Joe Biden is a relation between Y and Joe Biden which is *accidental* to him. But nothing which is accidental to him can be constitutive of his existence. Nothing which is accidental to him can explain what it is for Joe Biden to exist. For example, Joe Biden can cease to be contained by the White House or to appear in it, so that the White House ceases to have the property of containing him and being [Joe Biden]ish without it being the case that Joe Biden ceases to exist.

Even worse, suppose there is a restricted domain of objects Y whose having the property of containing him and being [Joe Biden]ish is constitutive of the existence of Joe Biden. In this case, we would have to revise our belief that Joe Biden is a substance, i.e., an object which exists *per se* and *in se*. We would have to give up the idea that Joe Biden is a substance because any object X, whose existence consists in something else Y's being [X]ish, exists only as an accident or modification of Y, i.e., as that by which Y is [X]ish.

A further problem is that even if the latter were the case, Y would have to be a subject which subsists, i.e., which does not exist in another as a subject, to make sense of the idea that Joe Biden exists as an accident or modification of Y. However, this possibility must be ruled out by the restricted-domain view because it claims that to exist is always to *exist in* something else. It follows that there must be a restricted domain of objects Z whose having the property of containing Y and being [Y]ish is constitutive of the existence of Y. Consequently, Y exists only as an accident or modification of a restricted domain of objects Z, i.e., as that by which Z is [Y]ish, and the problem repeats itself *ad infinitum*.

Thus, by the very form of its explanation, according to which to exist is always to *exist in*, the restricted-domain view is unable to explain what it is for an object which subsists—a substance— to exist, a substance being an object which does not exist in another as a subject like accidents or modifications of substances. In short, by the very form of its explanation of what it is for an object to exist, this view is unable to explain what it is for a subsistent object to exist.[12] In this way, the ontological status of subsistent objects is undermined, and the view is confronted with the problem that it is committed to an absurd 'accidentalization' of all subsistent objects.

Yet such a view is not only absurd, it cannot even be formulated in an intelligible way: A world full of accidents or modifications requires subjects which subsist and *of which* these accidents or modifications are accidents or modifications. But at the same time, the restricted-domain view must negate that such

12 This analysis has the additional advantage of shedding further light on the problem of the incoherence of the unrestricted-domain view (see 4.2.2). To make sense of the idea that the existence of an object X consists in Y's having the property of being [X]ish, Y needs not only exist but *subsist*, i.e., Y must exist in itself and not in another as a subject. But if to exist is always *to exist in* something else, it is impossible that Y exists in itself. Now, it is not only impossible for an unrestricted domain of objects to exist. It is impossible for any substance to exist because the very account of existence rules out the possibility that something exists in itself and not in another as a subject, i.e., the possibility that something subsists.

subjects exist because it is committed to the idea that to exist is always to *exist in* which is the way an accident or modification exists.

To evade these devastating consequences, the restricted-domain view must maintain that for objects like Joe Biden there is no restricted domain of objects Y whose having the property of containing him and being [Joe Biden]ish is constitutive of the existence of Joe Biden. The view is forced to admit that any restricted domain of objects Y's having the property of containing Joe Biden and being [Joe Biden]ish is not constitutive of Joe Biden's existence because Y's having the property of containing Joe Biden is a relation between Y and Joe Biden which is *accidental* to him. But if Y's having the property of containing Joe Biden and being [Joe Biden]ish is not constitutive of Joe Biden's existence, the restricted-domain view fails to explain what it is for Joe Biden to exist.

Aquinas's view solves this problem by rejecting the premise which gives rise to it, namely, the idea that the existence of an object X—a substance in his terminology—consists in something else Y's having a property F. According to Aquinas, what is constitutive of the existence of a substance is a substance's having *esse*. To be more precise, what is constitutive of the existence of a substance is the substance's metaphysical composition of a determinable and a determinant, namely, its essence or nature and its *esse*. Aquinas's view has no problem with acknowledging that a substance is always embedded in contexts which are accidental to it because the substance's embeddedness in any context is not supposed to explain what it is for the substance to exist.

Furthermore, Aquinas's view resists the attempt to identify and reduce existence with and to *existence-in*. To exist is not always to *exist-in*. As we have seen above (see 6.1), Aquinas distinguishes between the way substances and accidents, i.e., modifications of substances exist. Substances subsist, i.e., they have *esse per se* and *in se*, while accidents or modifications of substances exist only in the qualified sense of having *esse* in a substance as their subject. From Aquinas's point of view, it is not surprising that the restricted-domain view produces the problems it has because it attempts to explain what it is for a subsistent object to exist in a way which can at best explain what it is for a non-subsistent entity to exist.[13]

13 I do not address three further problems I have identified in the first part which are specific to the unrestricted-domain view and the restricted-domain view because it is easy to see that Aquinas's view can solve them in similar ways as it solves the four problems I have dealt with so far. First, Aquinas's view solves the problem of incoherence (see 4.2.2) because it can explain what it is for an object X to exist without having to introduce an existing unrestricted domain of objects, i.e., a UDO. A substance X exists in virtue of having *esse* and not in virtue of a UDO's having the property of containing X and being [X]ish. Second, Aquinas's view solves the problem that re-

7.2 Why Aquinas's View Can Solve the Problems of the Object Views

Up to this point, I have explained why Aquinas's view can solve the problems of the three views which are united by the idea that the existence of an object X consists in something else Y's having a property F. In this section, I will explain why Aquinas's view can solve the problems of the remaining two views which belong to the rival camp. What unites the two views which belong to this camp is the idea that the existence of an object X consists in X's having a property F.

7.2.1 Solution to a Vicious Circularity

The first problem with the non-reductive-object view is that its explanation of what it is for an object to exist is viciously circular (see 1.2.1). What generates this problem is the very form of the view's explanation of what it is for an object to exist: The existence of an object X consists in X's having the property of existing. This way of answering the nature question generates the problem of a vicious circularity because X can only have the *property* of existing if X exists. An object X as a pre-existing subject and possessor of the property of existing must be presupposed to make sense of the idea that X has the *property* of existing. In other words, to explain the existence of X in terms of X's having the property of existing is viciously circular because the existence of X is logically and ontologically prior to X's having any property. An account of what it is to have a property presupposes an account of what it is for that which is supposed to have a property of existing.

Aquinas's view solves this problem by rejecting the idea which gives rise to the problem. Existence—*esse* in Aquinas's terminology—is not a property of an object—a substance in Aquinas's terminology—which has it. Aquinas's view agrees with the non-reductive-object view that the existence of a substance X consists in X's having *esse*, i. e., existence or actuality. But Aquinas offers a very different explanation of how this 'having' works. He does not explain what it is to exist, i. e., to have *esse*, in terms of an account of what it is to

ductionism does not work (see 5.2.5) simply by employing an explanatory strategy which is non-reductive. According to Aquinas, *esse* is primitive and cannot be identified with and reduced to a property or form which something else has. Finally, Aquinas's view needs not solve the problem of determining what kind of entity a restricted domain of objects is (see 5.2.6) because its answer to the nature question works without the introduction of such entities.

have a property. Rather, Aquinas explains a substance's having *esse* in terms of participation (see 6.2).

A fact of particular importance is that he carefully distinguishes the mode or way *esse* is participated in by a substance from the way or mode an accidental form is participated in by a substance (see 6.2.2). *Esse* is not participated in as an accidental form which the substance has in addition to its essence or form. Rather, it is a substance's essence or form—and consequently the substance itself taken as what it is essentially—which participates in *esse*. What distinguishes the way *esse* is participated in by a substance from the way an accidental form is participated in by a substance is that only in the former case the essence or nature—and consequently the substance as what it is essentially—is determined with respect to its actuality. *Esse* determines a substance to be in actuality as what it is essentially precisely because it is the essence or form of a substance which participates in *esse* and not any accidental form which the substance has in addition to its essence or form.

Due to this careful distinction between the way *esse* is participated in by the essence or form of a substance and the way an accidental form is participated in by substance, Aquinas's view is able to explain what kind of metaphysical composition of a substance is required to explain what it is for a substance to exist. The existence of a substance does not result from the substance's metaphysical composition of a subject and *esse* as an accidental form which the substance has in addition to its essence or form. Rather, the existence of a substance results from the substance's metaphysical composition of that which participates and has *esse* (*quod est*)—the substance considered as what it is in virtue of its essence or nature alone in abstraction from its *esse*—and the *esse* it participates in and has and in virtue of which its essence or form, and consequently the substance itself, taken as what it is essentially, is actual (*quo est*).

From Aquinas's point of view, the view that the existence of a substance consists in that substance's having *esse* as a property rests on a valid intuition, namely, that a substance's having *esse* is best explained in terms of that substance's metaphysical composition. This non-reductive-object view only gets it wrong when it comes to the question of what *kind* of metaphysical composition of a substance is required to explain what it is for a substance to exist. Since the view starts with the assumption that existence or *esse* is a *property* of a substance, it inevitably ends up with the wrong kind of metaphysical composition, namely, a kind of subject-accident metaphysical composition which in turn makes its explanation of what it is for a substance to exist viciously circular because this kind of metaphysical composition requires a pre-existing subject, i.e., a possessor of the property of existence or *esse* which is actual.[14]

Aquinas's answer to the nature question is not viciously circular because the kind of metaphysical composition of a substance which is supposed to explain its existence does not presuppose a pre-existing subject, i.e., a possessor of *esse* which is actual. The essence or form of a substance has no actuality independently or separately from the *esse* it participates in. In the same way, the *esse* of a substance does not pre-exist independently or separately from the essence or nature of a substance which determines and limits it to the *esse* of a particular essence or nature. Aquinas's view is not that the existence of a substance results from a metaphysical composition of a somehow pre-existing essence or nature which is somehow actualized by *esse* which is added to it. Rather, the existence of a substance consists in its metaphysical composition of that which participates and has (*quod est*)—its essence or nature—and that which is participated in—*esse*—by which (*quo est*) the substance's essence or nature, and consequently the substance itself taken as what it is essentially, is actual. A view according to which you can have participation without something which participates and something which is participated is incoherent and does not even make sense.

In Aquinas's view, that which participates—the essence or nature of a substance—and that which is participated in—the *esse* of a substance—do not exist prior to the substance it composes but is *simultaneously* created together with the substance which results from their composition.[15] Again (see 6.1), it is

14 Jeremy Skrzypek brought to my attention the theoretical possibility that a non-reductive-object view could be combined with a bundle theory which makes use of a relational instead of a compositional explanatory strategy. Consequently, it could be argued that the problem of a vicious circularity does not arise because the point of a bundle theory is that there is no pre-existing subject. I am willing to grant that such a view is a theoretical possibility. Nevertheless, this view does not solve the problem of a vicious circularity. It just pushes the problem to another level where it resurfaces even worse. If objects are nothing more than bundles of properties, the properties of the bundle must exist prior to the bundle, and they must exist in their own right and not just as metaphysical parts or constituents of the object or bundle, because a compositional explanatory strategy is rejected by this view. But what is it for any property of a bundle to exist? As far as I can see, there are two possible answers. First, it could be argued that any property of a bundle possesses existence as a property, i.e., that any such property is the subject of the property of existence. But this answer does not work because it is again viciously circular. A property can only be the subject of the property of existence if it exists. Even worse, now the number of pre-existing subjects has been multiplied. Second, it could be argued that any property of a bundle stands in relation to the property of existence. But again, such a relation can only obtain if both relata exist, so that the problem of the vicious circularity is again not solved.

15 Cf. *QDP*, q. 3, a. 1, ad 17: "Ad decimumseptimum dicendum, quod Deus simul dans esse, producit id quod esse recipit: et sic non oportet quod agat ex aliquo praeexistenti."; ibid. ad 12: "Ad

important to note that Aquinas distinguishes between the metaphysical composite —the substance—as that which truly and properly exists because it has *esse per se* and *in se* and its metaphysical parts or components which exist only in and with the substance as that by which (*quo est*) the substance subsists in actuality as that which it is essentially.[16] For this reason, Aquinas's view is not vulnerable to the objection of ontological inflationism either. Substantial and accidental forms of substances are not entities in the same sense as the substances in which they inhere. Aquinas does not 'reify' substantial and accidental forms. They exist only in the qualified sense of that by which (*quo est*) a substance has substantial and accidental *esse*.

Why the kind of metaphysical composition which Aquinas proposes to explain what it is for a substance to exist does not presuppose a pre-existing subject, i.e., a possessor of *esse* which is actual, can also be seen from a slightly different angle by making use of the terminology of determinate, determinable, and determinant. Let us begin with the participation of a substance in an accidental form, for example a banana which participates in yellowness. In this case, the determinate is the yellow banana. The determinable is the banana which participates and has yellowness but considered in abstraction from the yellowness it has and participates in. The determinant is yellowness which is participated in, and which determines the banana to be yellow in actuality. This kind of metaphysical participation and composition clearly presupposes that the banana as a subject and possessor of yellowness exists prior to its participation in and composition with yellowness. The accidental form of yellowness can inhere in the banana as that by which the banana is yellow only if the banana exists. In other

---

duodecimum dicendum, quod neque materia neque forma neque accidens proprie dicuntur fieri; sed id quod fit est res subsistens. Cum enim fieri terminetur ad esse, proprie ei convenit fieri cui convenit per se esse, scilicet rei subsistenti: unde neque materia neque forma neque accidens proprie dicuntur creari, sed concreari. Proprie autem creatur res subsistens, quaecumque sit; ibid., corp.: Ipse autem Deus e contrario est totaliter actus,—et in comparatione sui, quia est actus purus non habens potentiam permixtam—et in comparatione rerum quae sunt in actu, quia in eo est omnium entium origo; unde per suam actionem producit totum ens subsistens, nullo praesupposito, utpote qui est totius esse principium, et secundum se totum. Et propter hoc ex nihilo aliquid facere potest; et haec eius actio vocatur creatio."

16 Cf. *QQ*, IX, q. 2, a. 2, corp.: "Sed hoc esse attribuitur alicui rei dupliciter. Uno modo, sicut ei quod proprie et vere habet esse vel est; et sic attribuitur soli substantiae per se subsistenti, unde quod vere est dicitur substancia in I Physicorum. Omnibus vero quae non per se subsistunt sed in alio et cum alio, sive sint accidentia sive formae substantiales aut quaelibet partes, non habent esse ita quod ipsa vere sint, sed attribuitur eis esse alio modo, id est ut quo aliquid est, sicut albedo dicitur esse, non quia ipsa in esse subsistat, sed quia ea aliquid habet esse album. Esse ergo proprie et vere non attribuitur nisi rei per se subsistenti."

words, the determinable is the banana considered in abstraction from yellowness but not considered in abstraction from the *esse* in virtue of which its essence or nature– and consequently the banana as what it is essentially–is actual.

The metaphysical composition from which the existence of a substance results is a different case in Aquinas's view because *esse* is not an accidental form which inheres in a subject. The determinant is the substance which exists. The determinant, i. e., the metaphysical composite, is the proper and true possessor of *esse* (*quod est*)$_1$. According to Aquinas's view, the substance's having *esse* or its participating in *esse* is explained in terms of its metaphysical composition of that which has and participates (*quod est*)$_2$ and that which it has and participates in (*quo est*), namely, its *esse*. What is important to note is that the substance, considered just as that which has and participates (*quod est*)$_2$ is not some kind of subject or possessor of *esse* which exists prior to the substance which properly and truly has *esse* (*quod est*)$_1$. Rather, it is what the substance (*quod est*)$_1$ is essentially but considered in abstraction from the *esse* it participates in as something outside of its essence or nature. Considered in this way, the substance (*quod est*)$_2$ is just a determinable with respect to *esse*. As a determinable, the substance, i. e., its essence or form, is not actual and does not exist prior to the substance which exists in actuality (*quod est*)$_1$ precisely because *esse* is outside of the essence or nature of the participating substance.

The abstraction helps us to grasp the determinability of the essence or form of a substance–and consequently of a substance itself taken as what it is essentially–with respect to its *esse* and to understand that a substance is related to its *esse* as a determinable to a determinant. The possibility of distinguishing between the substance as a determinate (*quod est*)$_1$ and a determinable (*quod est*)$_2$ is grounded in the fact that a substance participates in *esse* as something which is outside of its essence or nature.

It would be a mistake to conclude from this possibility of considering the essence or nature of a substance–and consequently the substance itself taken as what it is essentially–in abstraction from its *esse* that the substance considered in this way as a mere possessor of *esse* or participator in *esse* (*quod est*)$_2$ is a kind of pre-existing subject of *esse*. One who argues that the substance considered as a mere possessor of *esse* or participator in *esse* (*quod est*)$_2$ is a kind of pre-existing subject has not considered a substance as what it is in abstraction from its *esse*. In other words, they have not taken into account that *esse* is not included in the essence or nature of a substance which participates in *esse*.

7.2.2 Solution to the Problem That Existence Is Neither an Essential Nor an Accidental Property of an Object

The second problem with the non-reductive-object view is that existence is neither an essential nor an accidental property of an object which has existence (see 1.2.2). The solution which Aquinas's view has to offer to this problem is as simple as it is straightforward: It rejects the premise which gives rise to the problem, namely, that existence, i.e., *esse*, is a property of a substance. Only if one claims that *esse* determines a substance with respect to its actuality as a property determines a substance which has it, one encounters the problem that *esse* does not determine a substance as any other property which determines a substance that has it.

According to Aquinas's view, the solution to this problem is to stick with the claim that *esse* determines a substance with respect to its actuality but to reject the claim that *esse* is a property of a substance. *Esse* is not an essential property of a substance because *esse* is not included in the essence or nature of this substance. Rather, *esse* is participated in as something which is outside of the essence or nature of the participating substance. *Esse* is neither an accidental property nor form of a substance which the substance has in addition to its essence or form. A substance does not participate in *esse* as it participates in an accidental form because an accidental form that a substance participates in determines this substance in an accidental formal respect which *esse* does not. Furthermore, anything which is accidental to a substance cannot account for the substantial *esse* of a substance. Rather, *esse* determines the essence or nature of a substance which participates in it— and consequently the substance itself as what it is essentially—with respect to its actuality.

Thus, Aquinas's view has a way to explain how a substance has *esse* contingently without it being the case that a substance is accidentally related to the *esse* which determines the substance with respect to its actuality. Aquinas's decisive insight is that the essence, nature, or form of a substance participates in *esse*. In this way, it is possible for Aquinas to reconcile the view that a substance is an *ens* by participation with the view that a substance has substantial *esse* and subsists as what it is essentially, i.e., that a substance is an *ens* which exists *per se* and *in se*. As an *ens* by participation a substance only exists contingently in contrast to an *ens* which is an *ens* by its essence and which therefore exists necessarily. But a substance has substantial *esse*, i.e., it subsists as that which it is in actuality, precisely because it is its essence, nature, or form of a substance, and not any accidental form the substance has in addition to its essence or form, which participates in *esse*.

7.2.3 Solution to the Problem That What Is Supposed to Be Explained Is Not Explained

The third problem with the non-reductive-object view is that it does not explain what it is supposed to explain (see 1.2.3). An object's having the property of existing is supposed to explain what it is for an object to exist. The core idea of the non-reductive-object view is that existence is a property which determines the object which has it with respect to its actuality. To solve the problem that existence as a property does not determine the object which has it as an essential or accidental property determines its object, this view must argue that existence is a property *sui generis*. Therefore, this view must be amended to the view that the existence of an object X consists in X's having a property *sui generis*, namely, the property of existing.

Through this move, this view loses explanatory power, as I showed above in detail (see 1.2.3). What is supposed to do the explanatory work is the idea that an object is related to its existence in the way an object is related to a property which determines it. Pressed by the objection that this explanation does not work because existence does not determine its object like an accidental or essential property, this view is amended to the view that existence is a property *sui generis* of an object which determines its object with respect to its existence.

This move is unsatisfying because by qualifying existence as a property *sui generis* the non-reductive-object view leaves us where we started. The view that the existence of an object consists in that object's having a property *sui generis*, namely, the property of existing, which determines it with respect to its existence or actuality does not explain anything. It does not explain anything because it is left unexplained how an object is related to existence as a property *sui generis* of it and how existence as a property *sui generis* determines its object with respect to its existence or actuality.

Simply to reiterate that existence is a property *sui generis* which determines the object which has it with respect to its existence is of no help because it remains unexplained how a property *sui generis* determines its object. Something which needs to be explained, namely, how an object is related to its existence and how existence as a property determines its object, is explained with something which itself needs to be explained, namely, how an object is related to existence as a property *sui generis* and how existence as a property *sui generis* determines its object. Unless it is explained how a property *sui generis* determines its object and how such a property *sui generis* relates to an object which has it, nothing is explained by the view that for an object to exist is to have a property *sui generis*, namely, the property of existing.

Aquinas's view has a simple solution to this problem: It rejects the idea that a substance as what it is essentially is determined with respect to its actuality by a property it has. *Esse* or existence is not a property of a substance, not even a property *sui generis.*[17] According to Aquinas, *esse* determines a substance with respect to its actuality. But this determination does not work in the way any property determines a substance which has it. By making use of the ideas of metaphysical participation (see 6.2) and composition (6.3), Aquinas's view can explain what is supposed to be explained.

First, a substance is related to its *esse* as a determinable to a determinant. *Esse* is participated in by a substance as something which is outside of its essence or nature and as that by which (*quo est*) the essence, nature, or form of the participating substance—and consequently the substance itself taken as what it is essentially—subsists in actuality.

Second, Aquinas's view has a sophisticated way to explain how this determination works. A substance is a determinate which results from its metaphysical composition of a determinable and its *esse* as a determinant. Since a substance participates in *esse* as something which is not included in its essence or nature, a substance considered as what it is in virtue of its essence or nature alone, i.e., in abstraction from the *esse* it participates in, is just a determinable with respect to *esse*. Thus, in Aquinas's view, the existence of a substance consists in the substance's metaphysical composition of a determinable—its essence, nature, or form—and a determinant, namely, its *esse*.

7.2.4 Solution to the Three Remaining Problems

What remains to be explained is why Aquinas's view can solve the three problems faced by the reductive-object view. According to the latter view, existence is a property of an object but not a property *sui generis*. Rather, existence can be identified with and reduced to an object's property of being numerically identical with itself. As I have shown above in detail (see 2.2), such a view is confronted with three problems. First, the attempt to reduce existence to an object's prop-

---

17 Again, it could be objected that if the term 'property' is taken in a broad sense, *esse* is a property *sui generis* in Aquinas's view. This is true. But if *esse* is taken in this broad sense nothing is explained. In particular, it is not explained how *esse* belongs to a substance and how *esse* determines a substance with respect to its actuality. So, my emphasis here is that Aquinas rejects the idea that the characterization of *esse* as a property is helpful to answer the nature question. I am thankful to James Dominic Rooney and Jeremy Skrzypek for making me aware of the need to address this issue.

erty of being numerically identical with itself does not work because it fails to fulfill its own standards for a successful reduction (see 2.2.1). Second, if existence were identity, existence would be an essential property of an object and all existing objects would exist necessarily (see 2.2.2). Finally, if existence were identity, all objects identical with themselves would exist, which is obviously not the case (see 2.2.3).

Aquinas's view solves the first problem simply by negating the claim that existence or *esse* can be identified with and *reduced* to an object's property of being numerically identical with itself. A substance does not subsist in actuality as what it is essentially in virtue of being identical with itself. Its identity does not account for its existence or actuality. According to Aquinas's view, *esse* is primitive and cannot be identified with and reduced to anything else.[18] Consequently, there is no problem because there is nothing to reduce. On the contrary, the failure of the reductive-object view to identify and reduce existence with and to identity just confirms Aquinas's non-reductive view. If *esse* is primitive, any attempt to identify and reduce *esse* with and to something else must fail.

Aquinas's view solves the second problem by rejecting the premise that existence or *esse* can be identified with and reduced to an object's *property* of being identical with itself. The existence of a substance does not consist in that substance's having the property of being identical with itself. Consequently, Aquinas's view is not confronted with the problem that existence is an essential property of an object and that all existing objects exist necessarily. On the contrary, with the help of the ideas of metaphysical participation and composition, Aquinas is able to explain how a substance can exist contingently without it being accidentally related to its *esse*. A substance exists contingently and not necessarily because it is an *ens* by participation, i. e., a metaphysical composite of its essence or nature and its *esse*. Since *esse* is participated in by the essence, nature, or form of a substance, the substance is not accidentally related to *esse* but has substantial *esse*, i. e., it subsists as that which it is essentially in actuality.

Finally, Aquinas's view solves the third problem by rejecting the premise that existence or *esse* can be identified with and reduced to an object's property of *being (numerically) identical with itself*. From Aquinas's point of view, it is not surprising that van Inwagen's view has the problem that there are objects which have the property of being identical with themselves and yet do not exist (e. g., fictional or mythological objects like Sherlock Holmes or Pegasus). According to Aquinas, an object's property of being identical with itself is only a logical prop-

---

18 Cf. *SCG*, I, cap. 23: "Nihil enim est formalius aut simplicius quam *esse*."

erty and not a real property of an object which determines its object in an accidental or essential formal way.

To be more precise, identity is not a real relation but a logical relation, i.e., a relation which is a mere product of the human mind.[19] In order to grasp whether an object X is numerically identical with itself, the mind considers X as if it were two objects—X and Y—and asks whether X is numerically identical with Y.[20] In mind-independent reality there is of course no such relation because there is no Y which X is related to if X is numerically identical with itself. Therefore, the property or relation of identity between an object X and itself is a mere logical relation which follows upon our mode or way of understanding the numerical identity or numerical oneness of X. The important point is that since the relation of identity between an object and itself is a mere logical relation which follows upon our mode of understanding, this relation does not require that its *relata* exist, i.e., that the *relata* are actual *entia*.[21] Consequently, mere beings of reason (*entia rationes*) such as fictional or mythological entities (e.g., Sherlock Holmes or Pegasus) can also have the property of being identical with themselves.

Thus, Aquinas's view solves the problem that there are objects like Sherlock Holmes or Pegasus which have the property of being identical with themselves and yet do not exist by rejecting the idea which gives rise to this problem, namely, the idea that objects exist in virtue of being identical with themselves. Existence or *esse* cannot be reduced to the property of an object of being identical with itself. Identity is not actuality.

---

19 Cf. *QDV*, q. 1, a. 5, ad 15: "Unde inveniuntur quaedam relationes, quae nihil in rerum natura ponunt, sed in ratione tantum; quod quidem quadrupliciter contingit, ut ex dictis philosophi et Avicennae sumi potest. Uno modo, ut quando aliquid ad se ipsum refertur, ut cum dicitur idem eidem idem."

20 Cf. *QDP*, q. 7, a. 11, corp.: "Quandoque vero accipit unum ut duo, et intelligit ea cum quodam ordine: sicut cum dicitur aliquid esse idem sibi; et sic talis relatio est rationis tantum."

21 Cf. *QDP*, q. 7, a. 11, corp.: "Quandoque autem intellectus accipit aliqua duo ut entia, quorum alterum tantum vel neutrum est ens: sicut cum accipit duo futura, vel unum praesens et aliud futurum, et intelligit unum cum ordine ad aliud, dicens alterum esse prius altero; unde istae relationes sunt rationis tantum, utpote modum intelligendi consequentes."

Conclusion

In the previous chapter, I explained in detail why Aquinas's view can solve the problems of the views discussed in the first part of my book. What is shown thereby is that Aquinas's view can explain in a more satisfactory way what it is for an object to exist. By way of conclusion, I will argue that the contribution of Aquinas's view to the contemporary debate is even more far-reaching.

First, Aquinas's view and its answer to the nature question, i.e., its solution to the problem of explaining what it is for an object to exist, enables us to see why the whole contemporary debate has reached an impasse. It has come to this point because it has taken for granted a premise which Aquinas's view rejects. It is this rejection which allows Aquinas's view to have a new look at the problem and to formulate a more satisfactory answer to the nature question. Second, the contribution of Aquinas's view to the debate consists not in the reformulation or development of an already existing and established position. Rather, his view introduces a new explanatory strategy into the contemporary debate. Finally, Aquinas's explanatory strategy has the potential to integrate and synthesize the valid insights of the explanatory strategies it rejects.

Let us begin with the first contribution of Aquinas's view to the contemporary debate. Aquinas's view on what it is for an object to exist enables us to see why the whole contemporary debate about the nature question has reached an impasse. It had to develop into this state because the debate has taken for granted a premise which Aquinas's view rejects. To identify this premise, let us consider what is common to all the views I have discussed in the first part.

Despite all the differences between the five views, their different explanations of what it is for an object to exist have a common form: The existence of an object X consists in something Y's having a property F. Based on this common form, it can be nicely explained where the frontlines run in the debate.

The first question which divides the parties in the debate is the question whether X and Y are numerically identical. Views which respond positively to this question maintain that the existence of an object X consists in that object X's having a property F. According to these views, *objects* have a property F which accounts for their existence. The views which pertain to this camp are the non-reductive and the reductive-object view (see 1 and 2).

The remaining three views I discussed in the first part deny that X and Y are numerically identical. They all maintain that the existence of an object X consists in *something else Y's* having a property F. Views in this camp distinguish themselves with respect to what Y is and what F is. Y is either a property (see 3), or an unrestricted domain of objects (see 4), or a restricted domain of objects (see 5). F

https://doi.org/10.1515/9783110979879-011

is either a property's property of being instantiated (see 3), or the property of an unrestricted or restricted domain of objects of containing X and being [X]ish (see 4 and 5).

The second question which divides the parties in the debate is the question whether the property F is existence. The only view which responds positively to this question is the non-reductive-object view according to which the existence of an object X consists in X's having the property of existing (see 1). All the other discussed views negate this. Instead, they claim that existence can be identified with and reduced to another property. These views only disagree about what the property F is to which existence is reduced to. The systematic options I have discussed are that existence is an object's property of being identical with itself (see 2), that existence is a property's property of being instantiated (see 3), and that existence is the property of an unrestricted or restricted domain of objects of containing X and being [X]ish (see 4 and 5). In this case, the frontline runs between a non-reductive answer to the nature question according to which the existence of an object X consists in X's having a property *sui generis*, namely, the property of existing, and reductive answers to the nature question according to which the existence of an object X consists in something's having a property other than existence. In other words, the latter views maintain that existence can be identified with and reduced to another property of an object, a property, or a domain of objects.

What is taken for granted by all the discussed views is the following: The existence of an object X consists in something's *having a property F. Having a property F* plays a decisive role in all the different ways in which the considered views try to explain what it is for an object X to exist. It is precisely this premise which is rejected by Aquinas's view. In other words, what is distinctive about Aquinas's view is that it answers the nature question without making use of the idea that it is something's *having a property F* which accounts for the existence of an object X.

According to Aquinas, the existence of an object X consists in X's having *esse* (see 6.1). But Aquinas's view is not a variant or version of the non-reductive-object view because it rejects the claim that *esse* or existence is a property or accidental form of X. Rather, *esse* is primitive and an object X's having *esse* is explained in a different way by Aquinas. Thus, the idea that *having a property F* plays a decisive role in an explanation of what the existence of an object X consists in is rejected root and branch by Aquinas's view. It is this rejection of a premise which so far has been taken for granted which makes an important contribution to the contemporary debate.

It makes us aware that a third question must be asked which has not been asked so far: Does the existence of an object X consists in something's *having a property F?* If this third question is asked, we get an interesting new division of the contemporary debate about the nature question. All the views I have discussed in the first part respond positively to this question. Only Aquinas's view denies it. Asking this neglected third question enables us to see why the whole contemporary debate has reached an impasse. The debate has reached an impasse because the discussed non-Thomistic answers to the nature question wrongly assume that something's *having a property F* must play a role in an explanation for what it is for an object X to exist.

If one starts with the idea that the existence of an object X consists in something's having a property F, one ends up quite plausibly with a view which makes use of one of the two established strategies to explain what it is for something to have a property F. In other words, if something's having a property F plays a decisive role in the explanation of what it is for an object X to exist, one requires an account of what it is for something to *have* a property. There are basically two rival accounts or explanatory strategies available which try to explain how this *having* is supposed to work. In what follows, I will argue that the views discussed in the first part have the problems they have because they employ one of these two explanatory strategies to explain what it is for an object to exist.

The first explanatory strategy is usually called the 'constituent' strategy but for reasons which will immediately become clear, I prefer to call it the 'compositional' explanatory strategy. The core idea of this strategy is that something X has a property F in virtue of X being a metaphysical composite which has F as a metaphysical part.[1] Thus, X's having F is explained in terms of the metaphysical composition of X and F. For example, as we have seen above (see 6.1), Aquinas maintains that material substances are metaphysically composed of their substantial form, their prime matter, and a variety of accidental forms.[2] What explains a material substance's having an accidental property or form is that the relevant accidental form inheres in the substance as a metaphysical part of it.

---

1 Cf. Koslicki, *Form, Matter, Substance*, 11; van Inwagen, "Against Ontological Structure," 50; William F. Vallicella, "Constituent Versus Relational Ontology," *Studia Neoaristotelica* 10, no. 1 (2013): 100 – 101; Michael J. Loux, "What Is Constituent Ontology?," in *Metaphysics: Aristotelian, Scholastic, Analytic*, ed. Lukas Novak, et al. (Heusenstamm: ontos verlag, 2012), 44 – 45.

2 This claim is not accepted by all contemporary interpreters of Aquinas. Some maintain that his view is that a material substance is only composed of prime matter and a substantial form, and that accidental forms do not enter into the composition of a substance. Thus, according to the latter interpretation of Aquinas's view, accidental forms are not metaphysical parts of a material substance which has them. For a good overview of the contemporary debate and a convincing argument in favor of the interpretation of Aquinas's view which I embrace in this inquiry, see Skrzypek, "Accidental Forms as Metaphysical Parts."

For example, what explains a yellow banana's having yellowness is that yellowness inheres in the banana as an accidental metaphysical part of it. The yellow banana is a metaphysical composite of its substantial form, prime matter, and the form yellowness which the banana has in addition to its substantial form, and which is accidental to it. In other words, what accounts for the banana's yellowness is that the banana is metaphysically composed in the manner of subject-accident. The substantial form of the banana configures prime matter, and the resulting banana serves as the subject of the accidental form of yellowness which determines the actual banana with respect to its quality, i.e., its color.

The second explanatory strategy is usually called the 'relational' strategy. The employment of this strategy is often motivated by the rejection of the view that objects are metaphysical composites, i.e., that they have metaphysical parts, and the adoption of the view that objects are not metaphysically structured but simple.[3] If the view that objects are metaphysical composites which have metaphysical parts is rejected, a different account is required for what it is for something to have a property or to exhibit a certain accidental characteristic. A different explanation is needed for how the 'having' of a property is supposed to work.

The core idea of the relational explanatory strategy can be illustrated in this way: Let us suppose there exists a yellow banana and we agree that being yellow is accidental to the banana. Now, what explains the yellowness of the banana is not that the yellow banana is a metaphysical composite of a banana and the accidental form of yellowness, i.e., that yellowness inheres in the banana as a metaphysical part of it. Rather, what explains the yellowness of the banana is that the banana as a concrete particular and metaphysically simple object stands in an appropriate *relation* to yellowness as an abstract object which is *external* to it.[4] For example, it has been argued that the appropriate relation is a relation of instantiation, exemplification, or participation.[5] Thus, the banana is yellow or has yellowness in virtue of standing in a relation of instantiation, exemplification, or participation to yellowness which is external to it.[6]

---

3 Cf. Vallicella, "Constituent Versus Relational Ontology," 101; van Inwagen, "Against Ontological Structure," 51–54.

4 Cf. Vallicella, "Constituent Versus Relational Ontology," 101; van Inwagen, "Against Ontological Structure," 52; Bradley Rettler and Andrew M. Bailey, "Object," ed. Edward N. Zalta, Winter 2017 ed., *Stanford Encyclopedia of Philosophy* (2017), https://plato.stanford.edu/archives/win2017/entries/object/.

5 Cf. Koslicki, *Form, Matter, Substance*, 13; Rettler and Bailey, "Object."; Vallicella, "Constituent Versus Relational Ontology," 101; van Inwagen, "Against Ontological Structure," 58.

6 Cf. Koslicki, *Form, Matter, Substance*, 16.

The five views considered in the first part generate the problems they have because they employ one of these two strategies to explain what it is for an object to exist.[7] Let us begin with the property view, the unrestricted-domain view, and the restricted-domain view. What is common to these three views is the idea that the existence of an object X consists in something else Y's having a property F. Y's having a property F is supposed to explain the existence of X because F is a dyadic or two-place property, i.e., a relation, which obtains between Y and X. In other words, the existence of X or X's 'having' existence is explained in terms of a relation which obtains between X and something else Y which is external to X and not in terms of the metaphysical composition of X. Thus, the three views have in common that they employ a relational explanatory strategy.

According to the property view, the existence of an object X consists in the property of a property Y, which individuates X, of being instantiated. Thus, it is a relation of instantiation or exemplification which obtains between a property Y and an object X which is supposed to explain the existence of X. According to the unrestricted-domain view and the restricted-domain view, the existence of an object X consists in the property of an unrestricted or restricted domain of objects Y of containing X and being [X]ish. In this case, it is a relation of containment which obtains between a domain of objects Y and an object X which is supposed to explain the existence of X.

Whatever merits a relational explanatory strategy might have as a strategy to explain an object X's 'having' a property F, it clearly does not work as a strategy to explain what it is for an object X to exist. It is not difficult to see why such views, which maintain that the existence of an object X consists in its being related in the appropriate way to *something else* Y which is external to X, generate the identified problems. For example, whatever 'Y' stands for, it must itself exist and Y's existence needs to be explained to provide a satisfactory explanation for what it is for X to exist. As I have explained above in detail, it is this requirement

7 It could be objected that the reductive-object view employs neither a compositional nor a relational explanatory strategy. I grant that this view is a special case. Nevertheless, it could be argued that it does employ a relational explanatory strategy. It only gives up the claim that an object has existence in virtue of standing in relation to *something else* which is external to it. Rather, the reductive-object view claims that X has existence in virtue of standing in relation to *itself*, namely, the relation of numerical identity. It is this peculiarity of its relational strategy which explains why this view generates other problems than the views which argue that an object X exists in virtue of standing in relation to something else which is external to it. To keep things as simple as possible, I will bracket this special case and limit my argument to the views which maintain that an object X exists in virtue of standing in relation to something else which is external to it.

which produces the problem of a vicious circularity (see 7.1.1), a vicious infinite regress (see 7.1.2), and the problem of an ontological inflationism (see 7.1.3).

This leaves us with the non-reductive-object view. Now, since the core idea of this view is that the existence of an object X consists in X's having the property of existing, it is quite plausible that this view tries to explain X's having existence in the way it tries to explain X's having a property F, namely, by making use of a compositional explanatory strategy. The attempt to answer the nature question in this way generates the identified problems (see 1.2). For example, the explanation of the view of what it is for an object X to exist becomes viciously circular because the metaphysical composition of an object X and existence as an accidental property of it, which is supposed to explain X's having existence, presupposes the existence of X as a subject of F. The inherence of existence as a property in X does not explain but rather presupposes the existence of X in which it inheres. Furthermore, the subject-accident composition of an object cannot explain how an object's having existence works because an object can cease to be composed in this way without ceasing to have existence and to exist. Thus, if the metaphysical composition of an object accounts for the object's having existence, it must be a different kind of metaphysical composition.

Up to this point, we have seen that Aquinas's view rejects a premise which has been taken for granted in the contemporary debate about the nature question, namely, that the existence of an object X consists in something's *having a property F.* It is the rejection of this premise which allows Aquinas's view to formulate a more satisfactory answer to the nature question.

What Aquinas's view contributes to the contemporary debate is therefore not a reformulation or a development of an already existing and established position. Rather, his view introduces a new explanatory strategy into the debate which has not been considered so far precisely because none of the discussed views has questioned the idea that something's having a property F must figure into an explanation of what it is for an object to exist. To get a better grasp of the newness of Aquinas's view, I will briefly summarize how it distinguishes itself from all the other discussed views.

First, Aquinas's view distinguishes itself from the property view, the unrestricted-domain view, and the restricted-domain view through the rejection of the claim that the existence of an object X consists in something else Y's having a property F. Aquinas rejects the relational explanatory strategy which these views employ to explain what it is for an object to exist. The existence of X does not consist in a relation which obtains between X and something Y which is *external* to it. Rather, that which accounts for the existence of an object X is the existence or *esse* which the object has. In other words, the existence of an object X results from a relation which is *internal* to X because it is a compositional relation. In this respect, Aquinas's view is similar to the non-reductive and reductive-object view. But the similarity is restricted to the shared idea that that which accounts for the existence of an object X belongs to X and not to something else Y.

Second, what distinguishes Aquinas's view from the reductive-object view is that it rejects the reductionism of the latter view. According to Aquinas's view, *esse* is primitive. It cannot be identified with and reduced to an object's property of being numerically identical with itself or any other property of an object. The existence of an object X consists in X's having *esse* and not in X's having the property of being identical with itself. In this respect, Aquinas's view is similar to the non-reductive-object view according to which the existence of an object X consists in X's having a property *sui generis*, namely, the property of existing. But the similarity is restricted to the shared idea that that which accounts for the existence of an object X belongs to X as something which is primitive.

Third, Aquinas's view distinguishes itself from the non-reductive-object view by rejecting the claim that the existence of an object X consists in X's having a property *sui generis*, namely, the property of existing. According to Aquinas's view, *esse* is not an accident or accidental form which a substance has in addition to its essence, nature, or form.[8] Consequently, a substance's having *esse* cannot be explained in the way a substance's having an accidental form is explained. What Aquinas's view shares with the non-reductive-object view is the general idea that the existence of an object X, i.e., X's having *esse*, is best explained by X's metaphysical composition. In other words, Aquinas's view has in common with the non-reductive-object view that it employs a compositional strategy to explain what it is for an object X to exist. The existence of X results from its metaphysical composition.

It is important to note that Aquinas's view employs a new and different compositional explanatory strategy. According to Aquinas, *esse* is not an accidental form of a substance which the substance has in addition to its essence, nature,

---

8 It could be argued that Aquinas's view is a variant of the non-reductive-object view if this view employs the term 'property' in a broad sense, according to which anything which can be truly predicated of an object or attributed to an object is a property of that object. In other words, if everything which somehow 'belongs' to an object is a property of it, it seems that Aquinas would agree to say that *esse* is a property of an existing object. But the problem is that if the term 'property' is used in this broad sense, it remains unexplained how an object is related to *esse* as its property, i.e., how an object's 'having' *esse* works. So, in my opinion, it does not make much sense and is rather confusing to classify Aquinas's view as a variant of the non-reductive-object view, because the idea that *esse* is a property in this broad sense explains nothing and plays no role in his answer to the nature question. I am thankful to James Dominic Rooney and Jeremy Skrzypek for making me aware of the need to address this issue.

or form. Consequently, it makes no sense to try to explain a substance's having *esse* in terms of a substance's metaphysical composition of a subject and an accidental form or property. If one starts with the idea that *esse* is a property or accidental form of a substance and then tries to explain a substance's having *esse* with the help of a compositional explanatory strategy one inevitably gets onto the wrong track right from the beginning.

In Aquinas's view, a substance X's having *esse* is explained in terms of participation, i.e., a participatory relationship between its essence, nature, or form and *esse* (see 6.2), and this participatory relationship is in turn explained through the metaphysical composition of X (see 6.3). The participation of X in *esse* results from X's metaphysical composition of a determinable and a determinant. Since *esse* is participated in by a substance as something which is outside of its essence or nature, a substance considered as what it is in virtue of its essence or nature alone, i.e., in abstraction from its *esse*, is just a determinable with respect to its *esse*. It is the *esse* of a substance which determines the essence, nature, or form of the substance—and consequently the substance itself taken as what it is essentially—with respect to its actuality.

To sum up: The decisive contribution of Aquinas's view to the contemporary debate is that it disentangles the idea that existence or *esse* is a property or accidental form of an object from the idea that the metaphysical composition of an object accounts for an object's having *esse*. Aquinas rejects the former idea but sticks with the latter. In this way, Aquinas's view introduces a new compositional explanatory strategy into the contemporary debate which can answer the nature question in a more satisfactory way than the discussed rival views.

So far, I have shown that Aquinas's view can formulate a satisfying answer to the nature question by employing a new compositional explanatory strategy. Aquinas's explanatory strategy can be summarized in this way: The existence of a substance X consists in X's having *esse* by means of its essence or form (see 6.1). X's having *esse* by means of its essence or form is explained in terms of metaphysical participation (see 6.2). *Esse* is participated in by a substance as something which is intrinsic to it but outside of its essence or nature. It is the essence, nature, or form of a substance which participates in *esse*. For this reason, a substance has *esse* by means of its essence or form. And *esse* which is participated in determines the essence, nature, or form of a substance—and consequently the substance itself taken as what it is essentially—with respect to its actuality. In turn, the participation in *esse* of a substance is explained in terms of metaphysical composition. The substance is metaphysically composed of that which participates and that which is participated in which are related as potentiality to actuality. Thus, Aquinas's final answer to the nature question is that the existence of a substance X consists in X's metaphysical composition of a determinable—its essence, nature, or form—and a determinant, namely, its *esse* (see 6.3).

I would like to conclude my book by pointing out Aquinas's view's potential to integrate and synthesize the valid insights of the explanatory strategies it rejects. Let us begin with the compositional explanatory strategy. What Aquinas's view rejects is the idea that *esse* or existence is a property or accidental form of a substance which the substance has in addition to its essence or form. Consequently, Aquinas's view rejects a compositional strategy which tries to explain a substance's having *esse* in terms of the subject-accident composition of a substance. *Esse* is not participated in by a substance like an accidental form is participated in by a substance. And the existence of a substance does not result from its being metaphysically composed in the manner of subject and accident.

Nevertheless, Aquinas's view does not reject the compositional explanatory strategy root and branch. Rather, Aquinas's view integrates the valid insight of this explanatory strategy, namely, the idea that for a substance to exist is to be metaphysically composed. Metaphysical composition plays an important role in Aquinas's solution to the problem of explaining what the existence of a substance consists in and how a substance's *having esse* works.

Now, what about the relational explanatory strategy? I have argued above that Aquinas rejects this explanatory strategy. The existence of a substance X does not consist in a relation which obtains between X and something Y which is *external* to X. Rather, that which accounts for the existence of X is the existence or *esse* which X has. In other words, the existence of a substance X results from a relation which is *internal* to X because it is a compositional relation.

However, it is important to note that Aquinas does not reject the relational explanatory strategy root and branch. Aquinas's rejection of it must be qualified. The relational strategy plays no role in Aquinas's way to answer the nature or weak ground question which has been the topic of this book. Nevertheless, as I made clear in the introduction, the nature or weak ground question must be distinguished from the strong ground question. It is one thing to ask what it is for that which exists to exist, and it is another thing to ask *why* that which exists exists. We can use Aquinas's view on what it is for a substance to exist to illustrate why a satisfactory answer to the nature question does not automatically provide a satisfactory answer to the strong ground question.

Aquinas's answer to the nature question is that the existence of a substance X consists in X's having *esse* by means of its essence or form. And X's 'having' *esse* by means of its essence or form is explained in terms of metaphysical participation and composition. In this way, what is explained by Aquinas's answer to the nature question is how a substance's 'having' *esse* works, i.e., how a substance is related to its *esse*. What is not explained by this answer is in virtue of what X has its *esse*. Any substance has *esse* only contingently. But what accounts for the fact that a substance has *esse* it might as well not have? If a substance has *esse* and exists although it might not have *esse* and not exist, there must be an explanation for this fact. In other words, if the existence of a substance consists in a certain kind of metaphysical composition, there must be a different kind of explanation which explains the fact that a substance is metaphysically composed in this way.[9]

It is at this point that Aquinas makes use of a relational explanatory strategy. Aquinas employs a relational strategy to formulate a satisfactory answer to the strong ground question. An adequate reconstruction and detailed exposition of Aquinas's answer to this question is beyond the limits of this inquiry. But a brief sketch of the core idea of Aquinas's answer suffices to see how his view integrates a valid insight of the relational explanatory strategy. According to Aquinas, a substance has *esse* it could also not have in virtue of standing in relation to subsistent *esse* itself which is extrinsic to it, and which causes the substance to have the *esse* it has and in virtue of which it is actual as what it is essentially.[10]

Again, Aquinas describes this relationship in terms of metaphysical participation. Nonetheless, as I have explained above in detail (see 6.2), it is important to distinguish between two kinds of metaphysical participation: A participation by composition and a participation by similitude. While the former kind of participation plays a decisive role in his answer to the nature question, the latter kind of participation plays an important role in his answer to the strong ground question. In the case of a participation by composition something participates in something else which is *intrinsic* to it, while in the case of a participation by sim-

9 Cf. *SCG*, I, cap. 18: "Omnis compositio indiget aliquo componente: si enim compositio est, ex pluribus est; quae autem secundum se sunt plura, in unum non convenirent nisi ab aliquo componente unirentur. [...] Componens autem est causa efficiens compositi."

10 Cf. *DEE*, cap. 4: "Omne autem quod convenit alicui vel est causatum ex principiis naturae suae, sicut risibile in homine; vel advenit ab aliquo principio extrinseco, sicut lumen in aere ex influentia solis. Non autem potest esse quod ipsum esse sit causatum ab ipsa forma vel quiditate rei, dico sicut a causa efficiente, quia sic aliqua res esset sui ipsius causa et aliqua res se ipsam in esse produceret: quod est impossibile. Ergo oportet quod omnis talis res cuius esse est aliud quam natura sua habeat esse ab alio. Et quia omne quod est per aliud reducitur ad id quod est per se sicut ad causam primam, oportet quod sit aliqua res quae sit causa essendi omnibus rebus eo quod ipsa est esse tantum; alias iretur in infinitum in causis, cum omnis res quae non est esse tantum habeat causam sui esse, ut dictum est. [...] Omne autem quod recipit aliquid ab alio est in potentia respectu illius, et hoc quod receptum est in eo est actus eius; ergo oportet quod ipsa quiditas vel forma quae est intelligentia sit in potentia respectu esse quod a Deo recipit, et illud esse receptum est per modum actus."

ilitude, something participates in something else which is *extrinsic* to it. A substance participates in substent *esse* itself which is extrinsic to it as an effect participates in the power of its cause.[11] Subsistent *esse* itself as extrinsic and uncreated *esse* is the efficient and exemplary cause of the created *esse* which is intrinsic to a substance.[12] The *esse* which a substance has and which is intrinsic

---

11 Cf. *In BDH*, lect. 2, 24: "Et similiter etiam effectus dicitur participare suam causam, et praecipue quando non adaequat virtutem suae causae, puta si dicamus quod aer participat lucem solis quia non recipit eam in claritate qua est in sole."; *DSS*, cap. 3: "Omne autem participans aliquid accipit id quod participat ab eo a quo participat, et quantum ad hoc id a quo participat est causa ipsius: sicut aer habet lumen participatum a sole, quae est causa illuminationis ipsius."; *In CA*, lect. 4, 109: "Si autem aliquid sic haberet infinitam virtutem essendi quod non participaret esse ab alio, tunc esset solum infinitum; et tale est Deus, ut dicitur infra in XVI propositione. Sed, si sit aliquid quod habeat infinitam virtutem ad essendum secundum esse participatum ab alio, secundum hoc quod esse participat est finitum, quia quod participatur non recipitur in participante secundum totam suam infinitatem sed particulariter."; ibid., 146: "Ipsum autem esse participatum vocat finitum quia non participatur secundum totam infinitatem suae universalitatis sed secundum modum naturae participantis."; *SCG*, I, cap. 32: "De Deo autem nihil dicitur per participationem: nam omne quod participatur determinatur ad modum participati, et sic partialiter habetur et non secundum omnem perfectionis modum."; *In DDN*, cap. 5, lect. 1, 629: "Omnia autem alia, sicut superius dictum est, habent esse receptum et participatum et ideo non habent esse secundum totam virtutem essendi, sed solus Deus, qui est ipsum esse subsistens, secundum totam virtutem essendi, esse habet."

12 Cf. *QDV*, q. 21, a. 4, corp.: "Hoc enim Deum omnes de Deo loquentes intelligunt, quod est omnium principium effectivum, cum oporteat omnia entia ab uno primo ente effluere. [...] Unde si prima bonitas sit effectiva omnium bonorum, oportet quod similitudinem suam imprimat in rebus effectis; et sic unumquodque dicetur bonum sicut forma inhaerente per similitudinem summi boni sibi inditam, et ulterius per bonitatem primam, sicut per exemplar et effectivum omnis bonitatis creatae. Et quantum ad hoc opinio Platonis sustineri potest. Sic igitur dicimus secundum communem opinionem, quod omnia sunt bona creata bonitate formaliter sicut forma inhaerente, bonitate vero increata sicut forma exemplari."; *In DDN*, cap. 5, lect. 1, 630–631: "Et quod plus est, neque est, secundum scilicet quod significatur tempus praesens, quia eius esse tempore non mensuratur, sed ipse est esse existentibus, non quidem ita quod ipse Deus sit esse formale existentium, sed eo modo loquendi utitur quo Platonici utebantur qui esse separatum dicebant esse existentium, inquantum compositiva per participationem abstractorum participantur. [...] Deinde, cum dicit: resumentes etc., colligit ea quae dicta sunt et dicit resumendo quod omnia existentia et mensurae essendi habent esse a praeexistente; et omne aevum et tempus, quae sunt durationes, sunt ex ipso et ipse est principium effectivum et causa finalis omnis saeculi et temporis et cuiuslibet quocumque modo existentis; et iterum omnia ipso participant, sicut prima forma exemplari; et non solum est causa quantum ad fieri rerum, sed et quantum ad totum esse et durationem, quod manifestat cum dicit: et a nullo existentium recedit: aedificatore enim recedente, domus remanet, quia est causa domus quantum ad fieri et non quantum ad esse, sed si Deus ab effectu recederet, effectus non remaneret, quia est causa ipsius esse. Et quia causa praeeminet effectibus, ipse est ante omnia et omnia consistunt in ipso, Colos. I, sicut effectus virtute praeexistunt in causa; et universaliter, quidquid quocumque modo

to it bears a certain likeness or similitude to subsistent *esse* itself which is extrinsic to it precisely because the latter is the efficient and exemplary cause of the former. So, strictly speaking, a substance participates in subsistent *esse* only by a certain likeness or similitude which its intrinsic *esse* bears to subsistent *esse* itself as its cause.[13]

To sum up: In Aquinas's view, the participation (by similitude) of a substance in subsistent *esse* itself explains why a substance has (created) *esse* and exists although it could not have *esse* and not exist. Any *ens* which is an *ens* by participation has *esse* and exists because it stands in a causal relation to subsistent *esse* itself which is extrinsic to it and in virtue of which it has the *esse* it has which is intrinsic to it.[14]

Aquinas's combination of a compositional strategy to answer the nature question with a relational strategy to answer the strong ground question has a very interesting implication for his view on substances. On the one hand, a substance subsists. It is an *ens* which has *esse per se* and *in se* (see 6.1). On the other hand, a participatory relationship to subsistent *esse* itself which is extrinsic to a substance is constitutive of the metaphysical composition of the substance and consequently of its existence. This participatory relationship is constitutive of the

---

est, praeexistit in primo ente, scilicet Deo et quantum ad esse quod habet in intellectu et quantum ad conservationem sui esse."

13 Cf. *QQ*, XII, q. 4, a. 1, corp.: "Sciendum ergo quod unumquodque quod est in potentia et in actu, fit actu per hoc quod participat actum superiorem; per hoc autem aliquid maxime fit actu, quod participat per similitudinem primum et purum actum; primus autem actus est esse subsistens per se; unde completionem unumquodque recipit per hoc quod participat esse. Unde esse est completivum omnis formae, quia per hoc completur quod habet esse, et habet esse cum est actu; et sic nulla forma est nisi per esse."

14 Cf. *ST*, I, q. 44, a. 1, corp.: "Respondeo dicendum quod necesse est dicere omne quod quocumque modo est, a Deo esse. Si enim aliquid invenitur in aliquo per participationem, necesse est quod causetur in ipso ab eo cui essentialiter convenit; sicut ferrum fit ignitum ab igne. Ostensum est autem supra, cum de divina simplicitate ageretur, quod Deus est ipsum esse per se subsistens. Et iterum ostensum est quod esse subsistens non potest esse nisi unum, sicut si albedo esset subsistens, non posset esse nisi una, cum albedines multiplicentur secundum recipientia. Relinquitur ergo quod omnia alia a Deo non sint suum esse, sed participant esse. Necesse est igitur omnia quae diversificantur secundum diversam participationem essendi, ut sint perfectius vel minus perfecte, causari ab uno primo ente, quod perfectissime est."; ibid., ad. 1: "Ad primum ergo dicendum quod, licet habitudo ad causam non intret definitionem entis quod est causatum, tamen sequitur ad ea qua sunt de eius ratione, quia ex hoc quod aliquid per participationem est ens, sequitur quod sit causatum ab alio. Unde huiusmodi ens non potest esse, quin sit causatum; sicut nec homo, quin sit risibile. Sed quia esse causatum non est de ratione entis simpliciter, propter hoc invenitur aliquod ens non causatum."; *SCG*, II, cap. 15: "Quod per essentiam dicitur, est causa omnium quae per participationem dicuntur: sicut ignis est causa omnium ignitorum inquantum huiusmodi."

existence of a substance because subsistent *esse* itself is the extrinsic source or cause from which the essence or form of a substance continuously receives its *esse* in virtue of which the substance subsists as what it is essentially.[15] Thus, what accounts for the existence of a substance is its essence or form because it is that by which a substance has its substantial *esse*. But the essence or form of a substance can give *esse* to a substance only because the essence or form as that which has *esse* participates in (a likeness of) subsistent *esse* itself and continuously receives *esse* from it as the source or cause of its *esse*.[16]

Thus, in Aquinas's view, there is a constitutive relationality of that which subsists and this relationality does not undermine the ontological status of substances as *entia* which subsist, i.e., which exist in their own right (see 7.1.5). Thus, according to the Thomistic synthesis, the structure of a substance as an *ens* by participation is indissolubly dyadic: It *subsists* as something which is constitutively *related* to something else, namely, to subsistent *esse* itself.[17]

15 Cf. *ST*, I, q. 8, a. 1, corp.: "Cum autem Deus sit ipsum esse per suam essentiam, oportet quod esse creatum sit proprius effectus eius; sicut ignire est proprius effectus ignis. Hunc autem effectum causat Deus in rebus, non solum quando primo esse incipiunt, sed quandiu in esse conservantur; sicut lumen causatur in aere a sole quandiu aer illuminatus manet. Quandiu igitur res habet esse, tandiu oportet quod Deus adsit ei, secundum modum quo esse habet. Esse autem est illud quod est magis intimum cuilibet, et quod profundius omnibus inest, cum sit formale respectu omnium quae in re sunt, ut ex supra dictis patet. Unde oportet quod Deus sit in omnibus rebus, et intime."; *QDP*, q. 5, a. 1, corp.: "Respondeo dicendum quod absque omni dubio concedendum est, quod res conservantur in esse a Deo, et quod in momento in nihilum redigerentur, cum a Deo desererentur. [...] Unde sequitur quod divina operatione cessante, omnes res eodem momento in nihilum deciderent, sicut auctoritatibus est probatum in argumentis sed contra."; ibid., ad 8: "Ad octavum dicendum, quod creatura deficeret divina actione cessante, non propter contrarium quod sit in materia, quia ipsum etiam cum materia cessaret: sed propter hoc quod creatura est ex nihilo; non propter hoc quod nihilum aliquid ageret ad corruptionem, sed non ageret ad conservationem."

16 Cf. *QDP*, q. 5, a. 1, ad 3: "Ad tertium dicendum, quod, quamdiu principia essentialia rerum sunt, tamdiu res conservantur in esse; sed et ipsa rerum principia esse desinerent, divina actione cessante."; ibid., ad 10: "Ad decimum dicendum, quod etiam praedicta actione cessante, forma deficeret, unde non posset esse essendi principium."

17 Cf. Clarke, *Explorations*, 108, 119–120; Schindler, "Norris Clarke on Person, Being, and St. Thomas."

Bibliography

A Thomas Aquinas's Works

De ente et essentia (*DEE*) Sancti Thomae Aquinatis doctoris angelici Opera omnia iussu Leonis XIII. P.M. edita, cura et studio fratrum praedicatorum, Roma 1882 ss., vol. 43: *De principiis naturae ad fratrem Sylvestrum. De aeternitate mundi. De motu cordis ad magistrum Philippum de Castro Caeli. De mixtione elementorum ad magistrum Philippum de Castro Caeli. De operationibus occultis naturae ad quondam militem ultramontanum. De iudiciis astrorum. De sortibus ad dominum Iacobum de Tonengo. De unitate intellectus contra Averroistas. De ente et essentia.* Edited by H.-F. Dondaine. Rome: Editori di San Tommaso, 1976.

De principiis naturae (*DPN*) Sancti Thomae Aquinatis doctoris angelici Opera omnia iussu Leonis XIII. P.M. edita, cura et studio fratrum praedicatorum, Roma 1882 ss., vol. 43: *De principiis naturae ad fratrem Sylvestrum. De aeternitate mundi. De motu cordis ad magistrum Philippum de Castro Caeli. De mixtione elementorum ad magistrum Philippum de Castro Caeli. De operationibus occultis naturae ad quondam militem ultramontanum. De iudiciis astrorum. De sortibus ad dominum Iacobum de Tonengo. De unitate intellectus contra Averroistas. De ente et essentia.* Edited by H.-F. Dondaine. Rome: Editori di San Tommaso, 1976.

De substantiis separatis (*DSS*) Sancti Thomae Aquinatis doctoris angelici Opera omnia iussu Leonis XIII. P.M. edita, cura et studio fratrum praedicatorum, Roma 1882 ss., vol. 40: *Contra errores Graecorum ad Urbanum papam. De rationibus fidei ad Cantorem Antiochenum. De forma absolutionis paenitentiae sacramentalis ad Magistrum Ordinis. De substantiis separatis ad fratrem Raynaldum de Piperno. Expositio super primam et secundam Decretalem ad archidiaconum Tudertinum.* Edited by H.-F. Dondaine. Rome: Ad Sanctae Sabinae, 1969.

Expositio in librum Boethii De hebdomadibus (*In BDH*) Sancti Thomae Aquinatis doctoris angelici Opera omnia iussu Leonis XIII. P.M. edita, cura et studio fratrum praedicatorum, Roma 1882 ss., vol. 50: *Super Boetium De Trinitate.* Edited by P.-M. J. Gils; Expositio libri Boetii De ebdomadibus. Edited by L.-J. Bataillon and C. A. Grassi, Leonine Commision. Rome-Paris: Cerf, 1992.

Expositio super librum Boethii De trinitate (*In BDT*) Sancti Thomae Aquinatis doctoris angelici Opera omnia iussu Leonis XIII. P.M. edita, cura et studio fratrum praedicatorum, Roma 1882 ss., vol. 50: *Super Boetium De Trinitate.* Edited by P.-M. J.

https://doi.org/10.1515/9783110979879-012

	Gils; *Expositio libri Boetii De ebdomadibus.* Edited by L.-J. Bataillon and C. A. Grassi, Leonine Commision. Rome-Paris: Cerf, 1992.
Expositio super librum De causis (*In CA*)	*Super librum De causis expositio.* Edited by H.-D. Saffrey, Fribourg-Louvain: Société Philosophique-Nauwelaerts, 1954.
Expositio super librum Dionysii De divinis nominibus (*In DDN*)	*In librum Beati Dionysii De divinis nominibus expositio.* Edited by C. Pera, with historical introduction and doctrinal synthesis by P. Caramello and C. Mazzantini. Turin-Rome: Marletti, 1950.
Sententia super Metaphysicam (*In Meta*)	*In duodecim libros Metaphysicorum Aristotelis expositio.* Edited by M.-R. Cathala and R. M. Spiazzi. Turin-Rome: Marietti, 1950.
Sententia super Physicam (*In Phy*)	Sancti Thomae Aquinatis doctoris angelici Opera omnia iussu Leonis XIII. P.M. edita, cura et studio fratrum praedicatorum, Roma 1882 ss., vol. 2: *Commentaria in octo libros Physicorum Aristotelis.* Rome: Ex Typographia Polyglotta S. C. de Propaganda Fide, 1884.
Quaestio disputata de anima (*QDA*)	Sancti Thomae Aquinatis doctoris angelici Opera omnia iussu Leonis XIII. P.M. edita, cura et studio fratrum praedicatorum, Roma 1882 ss., vol. 24/1: *Quaestiones disputatae de anima.* Edited by B.-C. Bazán, Leonine Commission. Rome-Paris: Cerf, 1996.
Quaestiones disputatae de potentia (*QDP*)	*Quaestiones disputatae de potentia.* In S. Thomae Aquinatis, *Quaestiones disputae.* 2 vols. Edited by P. M. Pession. Turin-Rome: Marietti, 1965, 1–276.
Quaestio disputata de spiritualibus creaturis (*QDSC*)	Sancti Thomae Aquinatis doctoris angelici Opera omnia iussu Leonis XIII. P.M. edita, cura et studio fratrum praedicatorum, Roma 1882 ss., vol. 24/2: *Quaestio disputata de spiritualibus creaturis.* Edited by J. Cos, Leonine Commission. Rome-Paris: Cerf, 2000.
Quaestio disputata de unione verbi incarnati (*QDUVI*)	*Quaestio disputata de unione verbi incarnati.* In S. Thomae Aquinatis, *Quaestiones disputae.* 2 vols. Edited by P. Bazzi et al. Turin-Rome: Marietti, 1965, 367–415.
Quaestiones disputatae de veritate (*QDV*)	Sancti Thomae Aquinatis doctoris angelici Opera omnia iussu Leonis XIII. P.M. edita, cura et studio fratrum praedicatorum, Roma 1882 ss., vol. 22: *Quaestiones disputatae de veritate.* 3 vols. Edited by A. Dondaine. Rome: Editori di San Tommaso, 1970–76.
Quaestiones quodlibetales (*QQ*)	Sancti Thomae Aquinatis doctoris angelici Opera omnia iussu Leonis XIII. P.M. edita, cura et studio fratrum praedicatorum, Roma 1882 ss., vol. 25: *Quaestiones de quodlibet.* Edited by R.-A. Gauthier, Leonine Commission. Rome-Paris: Cerf, 1996.

Summa contra gentiles (*SCG*) — *Liber de veritate catholicae Fidei contra errores infidelium seu Summa contra Gentiles: Textus Leoninus diligenter recognitus.* 3 vols. Edited by P. Marc et al. Turin-Rome: Marietti, 1961–67.

Summa theologiae (*ST*) — Sancti Thomae Aquinatis doctoris angelici Opera omnia iussu Leonis XIII. P.M. edita, cura et studio fratrum praedicatorum, Roma 1882 ss., vol. 4–12: *Summa theologiae cum Supplemento et commentariis Caietani.* Rome: Ex Typographia Polyglotta S. C. de Propaganda Fide, 1888–1906.

B Other Cited Literature

Anscombe, G.E.M., and Peter T. Geach. *Three Philosophers: Aristotle, Aquinas, Frege.* Oxford: Basil Blackwell, 1961.

Bathen, Norbert. *Thomistische Ontologie und Sprachanalyse.* Freiburg im Breisgau; München: Verlag Karl Alber, 1988.

Benoist, Jocelyn. *L'Adresse du Réel.* Paris: Librairie Philosophique J. Vrin, 2017.

Berto, Francesco. *Existence as a Real Property: The Ontology of Meinongianism.* Dordrecht; London: Springer, 2013.

Berto, Francesco, and Matteo Plebani. *Ontology and Metaontology: A Contemporary Guide.* London: Bloomsbury Academic, 2015.

Bliss, Ricki, and J. T. M. Miller, eds. *The Routledge Handbook of Metametaphysics.* London: Routledge, 2020.

Boghossian, Paul A. *Fear of Knowledge: Against Relativism and Constructivism.* Oxford: Clarendon Press, 2006.

Brock, Stephen L. "Harmonizing Plato and Aristotle on *Esse:* Thomas Aquinas and the *De hebdomadibus.*" *Nova et Vetera* 5, no. 3 (2007): 465–494.

Brown, Christopher M. *Aquinas and the Ship of Theseus: Solving Puzzles About Material Objects.* London: Continuum, 2005.

Buchheim, Thomas, and Markus Gabriel, eds. *Neutraler Realismus.* Freiburg im Breisgau: Verlag Karl Alber, 2016.

Cameron, Ross P. "Critical Study of Kris McDaniel's *The Fragmentation of Being.*" *Res Philosophica* 95, no. 4 (2018): 785–795.

Caputo, John D. *Heidegger and Aquinas: An Essay on Overcoming Metaphysics.* New York: Fordham University Press, 1982.

Chalmers, David J., David Manley, and Ryan Wasserman, eds. *Metametaphysics: New Essays on the Foundations of Ontology.* Oxford: Clarendon Press, 2009.

Clarke, William N. *Explorations in Metaphysics: Being-God-Person.* Notre Dame, Indiana: University of Notre Dame Press, 1994.

Clarke, William N. "The Meaning of Participation in St. Thomas." *Proceedings of the American Catholic Philosophical Association* 26 (1952): 147–157.

Clarke, William N. *The One and the Many: A Contemporary Thomistic Metaphysics.* Notre Dame, Indiana: University of Notre Dame Press, 2001.

Davison, Andrew. *Participation in God: A Study in Christian Doctrine and Metaphysics.* Cambridge: Cambridge University Press, 2019.

De Haan, Daniel. "Aquinas on *Actus Essendi* and the Second Mode of Participation." *The Thomist* 82, no. 4 (2018): 573–609.
Doolan, Gregory T. "Aquinas on *Esse Subsistens* and the Third Mode of Participation." *The Thomist* 82, no. 4 (2018): 611–642.
Doolan, Gregory T. *Aquinas on the Divine Ideas as Exemplar Causes*. Washington, D.C.: The Catholic University of America Press, 2008.
Fabro, Cornelio. "The Intensive Hermeneutics of Thomistic Philosophy: The Notion of Participation." *The Review of Metaphysics* 27 (1974): 449–491.
Fabro, Cornelio. *Partecipazione e causalità secondo S. Tommaso D'Aquino*. 2 ed. Turin: Società Editrice Internazionale, 1960. 1939.
Fine, Kit. "Essence and Modality: The Second *Philosophical Perspectives* Lecture." *Philosophical Perspectives* 8, no. Logic and Language (1994): 1–16.
Frost, Gloria. "Aquinas on the Intension and Remission of Accidental Forms." *Oxford Studies in Medieval Philosophy* 7 (2019): 115–146.
Gabriel, Markus, ed. *Der Neue Realismus*. Berlin: Suhrkamp, 2014.
Gabriel, Markus. "Einleitung." In *Der Neue Realismus*, edited by Markus Gabriel, 8–16. Berlin: Suhrkamp, 2014.
Gabriel, Markus. "Existenz, realistisch gedacht." In *Der Neue Realismus*, edited by Markus Gabriel, 171–199. Berlin: Suhrkamp, 2014.
Gabriel, Markus. *Fields of Sense: A New Realist Ontology*. Edinburgh: Edinburgh University Press, 2015.
Gabriel, Markus. *Fiktionen*. Berlin: Suhrkamp Verlag, 2020.
Gabriel, Markus. "Für einen nicht-naturalistischen Realismus." In *Seien wir realistisch. Neue Realismen und Dokumentarismen in Philosophie und Kunst*, edited by Magdalena Marszalek and Dieter Mersch, 59–88. Zürich; Berlin: Diaphanes, 2016.
Gabriel, Markus. *I Am Not a Brain: Philosophy of Mind for the Twenty-First Century*. Cambridge: Polity Press, 2017.
Gabriel, Markus. "The Meaning of 'Existence' and the Contingency of Sense." *Frontiers of Philosophy in China* 9, no. 1 (2014): 109–129.
Gabriel, Markus. "Metaphysik oder Ontologie?". *Perspektiven der Philosophie: Neues Jahrbuch* 42 (2016): 73–93.
Gabriel, Markus. "Nachwort: Abgesang und Auftakt." In *Angst vor der Wahrheit: Ein Plädoyer gegen Relativismus und Konstruktivismus*, Paul A. Boghossian, 135–156. Berlin: Suhrkamp, 2013.
Gabriel, Markus. *Neo-Existentialism: How to Conceive of the Human Mind After Naturalism's Failure*. Cambridge: Polity, 2018.
Gabriel, Markus. "Neutral Realism." *The Monist* 98, no. 98 (2015): 181–196.
Gabriel, Markus. "Repliken auf Diehl/Rosefeldt, Hübner, Rödl und Stekeler-Weithofer." In *Jahrbuch-Kontroversen 2: Markus Gabriel – Neutraler Realismus*, edited by Thomas Buchheim, 165–222. Freiburg; München: Verlag Karl Alber, 2016.
Gabriel, Markus. *Sinn und Existenz: Eine realistische Ontologie*. Berlin: Suhrkamp, 2016.
Gabriel, Markus. "Sinnfeldontologie oder reformierte Metaphysik? Replik auf Le Moli." *Perspektiven der Philosophie: Neues Jahrbuch* 42 (2016): 110–125.
Gabriel, Markus. *Warum es die Welt nicht gibt*. Berlin: Ullstein, 2013.
Gabriel, Markus. *Why the World Does Not Exist*. Cambridge: Polity Press, 2015. 2013.

Gaitsch, Peter, Sandra Lehmann, and Philipp Schmidt, eds. *Eine Diskussion mit Markus Gabriel: Phänomenologische Positionen zum Neuen Realismus.* Wien: Verlag Turia + Kant, 2017.

Galluzzo, Gabriele. "Aquinas on Mental Being." *Quaestio* 2010, no. 10 (2010): 83–97.

Galluzzo, Gabriele, and Riccardo Chiaradonna. *Universals in Ancient Philosophy.* Pisa: Edizioni della Normale, 2013.

Galluzzo, Gabriele, and Michael J. Loux. *The Problem of Universals in Contemporary Philosophy.* Cambridge: Cambridge University Press, 2015.

Geiger, Louis Bertrand. *La participation dans la philosophie de s. Thomas d'Aquin.* 2 ed. Paris: Librairie Philosophique J. Vrin, 1953. 1942.

Gibson, Quentin. *The Existence Principle.* Dordrecht; Boston; London: Kluwer Academic Publishers, 1998.

Gilson, Étienne. *Being and Some Philosophers.* 2 ed. Toronto: Pontifical Institute of Mediaeval Studies, 1952.

Goldschmidt, Tyron. *The Puzzle of Existence: Why Is There Something Rather Than Nothing?* New York: Routledge, 2013.

Grossmann, Reinhardt. *The Existence of the World: An Introduction to Ontology.* London: Routledge, 1992.

Hübner, Johannes. *Einführung in die theoretische Philosophie.* Stuttgart: Verlag J.B. Metzler, 2015.

Hübner, Johannes. *Komplexe Substanzen.* Berlin; New York: Walter de Gruyter, 2007.

Ishii, Robertson, Teresa Atkins, and Philip Atkins. "Essential vs. Accidental Properties." In *The Stanford Encyclopedia of Philosophy,* edited by Edward N. Zalta 2020. https://plato.stanford.edu/archives/win2020/entries/essential-accidental/.

Kanzian, Christian. "Sein als Existenz." In *Handbuch Ontologie*, edited by Jan Urbich and Jörg Zimmer, 351–357. Berlin: Springer-Verlag, 2020.

Keller, Albert. *Sein oder Existenz?* Pullach bei München: Berchmanskolleg Verlag, 1968.

Kenny, Anthony. *Aquinas on Being.* Oxford; New York: Oxford University Press, 2002.

Kerr, Gaven. *Aquinas and the Metaphysics of Creation.* New York: Oxford University Press, 2019.

Kerr, Gaven. *Aquinas's Way to God: The Proof in De Ente et Essentia.* Oxford: Oxford University Press, 2015.

Kerr, Gaven. "The Meaning of *Ens Commune* in the Thought of Thomas Aquinas." *Yearbook of the Irish Philosophical Society* (2008): 32–60.

Kerr, Gaven. "Thomistic *Esse* and Analytical Philosophy." *International Philosophical Quarterly* 55, no. 1 (2015): 25–48.

Klima, Gyula. "Aquinas' Reception in Contemporary Metaphysics." In *The Oxford Handbook of the Reception of Aquinas*, edited by Matthew Levering and Marcus Plested, 565–580. Oxford: Oxford University Press, 2021.

Klima, Gyula. "The Changing Role of 'Entia Rationis' in Medieval Semantics and Ontology: A Comparative Study with a Reconstruction." *Synthese* 96, no. 1 (1993): 22–58.

Knasas, John F. X. *Being and Some Twentieth-Century Thomists.* New York: Fordham University Press, 2003.

Knasas, John F. X. *Thomistic Existentialism & Cosmological Reasoning.* Washington, D.C.: The Catholic University of America Press, 2019.

Korman, Daniel Z. *Objects: Nothing Out of the Ordinary.* Oxford: Oxford University Press, 2015.

Koslicki, Kathrin. *Form, Matter, Substance.* Oxford: Oxford University Press, 2018.

Kripke, Saul A. *Naming and Necessity.* Oxford: Blackwell, 1980.

Kripke, Saul. A. *Reference and Existence: The John Locke Lectures.* New York: Oxford University Press, 2013.

Lando, Giorgio. *Mereology: A Philosophical Introduction.* London; New York: Bloomsbury Academic, 2017.

LeNotre, Gaston G. "Determinate and Indeterminate Dimensions: Does Thomas Aquinas Change His Mind on Individuation?". *American Catholic Philosophical Quarterly* 94, no. 4 (2020): 503–546.

Li Vecchi, Joseph P., Frank Scalambrino, and David K. Kovacs. *The Philosophy of Being in the Analytic, Continental, and Thomistic Traditions: Divergence and Dialogue.* London: Bloomsbury Academic, 2020.

Loux, Michael J. "What Is Constituent Ontology?". In *Metaphysics: Aristotelian, Scholastic, Analytic*, edited by Lukas Novak, Daniel D. Novotny, Prokop Sousedik and David Svoboda, 43–57. Heusenstamm: ontos verlag, 2012.

Lowe, Edward J. "A Neo-Aristotelian Substance Ontology: Neither Relational Nor Constituent." In *Contemporary Aristotelian Metaphysics*, edited by Tuomas Tahko, 229–248. Cambridge: Cambridge University Press, 2012.

Luckner, Andreas, and Sebastian Ostritsch. *Existenz.* Berlin; Boston: Walter de Gruyter, 2018.

Luckner Andreas, and Sebastian Ostritsch, eds. *Philosophie der Existenz: Aktuelle Beiträge von der Ontologie bis zur Ethik.* Berlin: J.B. Metzler, 2019.

McDaniel, Kris. *The Fragmentation of Being.* Oxford: Oxford University Press, 2017.

McDaniel, Kris. *This Is Metaphysics: An Introduction.* Hoboken, NJ: Wiley-Blackwell, 2020.

McGinn, Colin. *Logical Properties: Identity, Existence, Predication, Necessity, Truth.* Oxford: Clarendon Press, 2000.

Miller, Barry. *The Fullness of Being: A New Paradigm for Existence.* Notre Dame, Indiana: University of Notre Dame Press, 2002.

Mitchell, Jason A. "Aquinas on *Esse Commune* and the First Mode of Participation." *The Thomist* 82, no. 4 (2018): 543–572.

Nelson, Michael. "Existence." In *The Stanford Encyclopedia of Philosophy,* edited by Edward N. Zalta 2020. https://plato.stanford.edu/archives/sum2020/entries/existence/.

Nevitt, Turner C. "How to Be an Analytic Existential Thomist." *The Thomist* 82, no. 3 (2018): 321–352.

Parsons, Terence. *Nonexistent Objects.* New Haven; London: Yale University Press, 1980.

Pasnau, Robert. *Metaphysical Themes, 1274–1671.* Oxford: Clarendon Press, 2011.

Perovic, Katarina. "Bradley's Regress." In *The Stanford Encyclopedia of Philosophy,* edited by Edward N. Zalta 2017. https://plato.stanford.edu/archives/win2017/entries/bradley-regress/.

Price, Huw. *Naturalism without Mirrors.* Oxford: Oxford University Press, 2011.

Priest, Graham. *Towards Non-Being: The Logic and Metaphysics of Intentionality.* 2 ed. Oxford: Oxford University Press, 2016.

Rettler, Bradley, and Andrew M. Bailey. "Object." In *Stanford Encyclopedia of Philosophy,* edited by Edward N. Zalta 2017. https://plato.stanford.edu/archives/win2017/entries/object/.

Rumfitt, Ian. "Contingent Existents." *Philosophy* 78, no. 4 (2003): 461 – 481.

Schindler, David L. . "Norris Clarke on Person, Being, and St. Thomas." *Communio* 20, no. 3 (1993): 580 – 592.

Schmidt, Andreas. "Ipsum Esse subsistens. Thomas von Aquin und die existierende Existenz." In *Philosophie der Existenz: Aktuelle Beiträge von der Ontologie bis zur Ethik*, edited by Andreas Luckner and Sebastian Ostritsch, 1 – 13. Berlin: J.B. Metzler, 2019.

Sider, Theodore. "Williamson's many necessary existents." *Analysis* 69, no. 2 (2009): 250 – 258.

Skrzypek, Jeremy W. "Accidental Forms as Metaphysical Parts of Material Substances in Aquinas's Ontology." *Oxford Studies in Medieval Philosophy* 7 (2019): 67 – 114.

Spade, Paul Vincent. *Five Texts on the Mediaeval Problem of Universals: Porphyry, Boethius, Abelard, Duns Scotus, Ockham.* Indianapolis; Cambridge: Hackett, 1994.

Spencer, Joshua. "Unnecessary existents." *Canadian Journal of Philosophy* 43, no. 5/6 (2013): 766 – 775.

Stump, Eleonore. *Aquinas.* London: Routledge, 2003.

Sylvan, Richard. *Exploring Meinong's Jungle and Beyond: An Investigation of Noneism and the Theory of Items.* Canberra: Australian National University, 1980.

Tahko, Tuomas E. *An Introduction to Metametaphysics.* Cambridge: Cambridge University Press, 2015.

Te Velde, Rudi A. *Participation and Substantiality in Thomas Aquinas.* Leiden; New York; Köln: E.J. Brill, 1995.

Textor, Mark. "States of Affairs." In *The Stanford Encyclopedia of Philosophy*, edited by Edward N. Zalta 2020. https://plato.stanford.edu/archives/sum2020/entries/states-of-affairs/.

Turner, Jason. "Ontological Pluralism." *The Journal of Philosophy* 107, no. 1 (2010): 5 – 34.

Vallicella, William F. "Constituent Versus Relational Ontology." *Studia Neoaristotelica* 10, no. 1 (2013): 99 – 115.

Vallicella, William F. "Existence: Two Dogmas of Analysis." In *Neo-Aristotelian Perspectives in Metaphysics*, edited by Daniel D. Novotny and Lukas Novak, 45 – 75. New York; London: Routledge, 2014.

Vallicella, William F. *A Paradigm Theory of Existence: Onto-Theology Vindicated.* Dordrecht: Kluwer Academic, 2002.

van Inwagen, Peter. "Against Ontological Structure." In *The Problem of Universals in Contemporary Philosophy*, edited by Gabriele Galluzzo and Michael J. Loux, 46 – 64. Cambridge: Cambridge University Press, 2015.

van Inwagen, Peter. "Being, Existence, and Ontological Commitment." In *Metametaphysics: New Essays on the Foundations of Ontology*, edited by David J. Chalmers, David Manley and Ryan Wasserman, 472 – 506. Oxford: Clarendon Press, 2009.

van Inwagen, Peter. "McGinn on Existence." *The Philosophical Quarterly* 58, no. 230 (2008): 36 – 58.

van Inwagen, Peter. *Metaphysics.* 4 ed. Boulder, CO: Westview Press, 2014.

Ventimiglia, Giovanni. *Aquinas after Frege.* Cham, Switzerland: Palgrave Macmillan, 2020.

Wasserman, Gloria. "Thomas Aquinas on Truths About Nonbeings." *Proceedings of the American Catholic Philosophical Association* 80 (2006): 101 – 113.

Williams, C.J.F. *What is Existence?* Oxford: Clarendon Press, 1981.

Williamson, Timothy. "Necessary Existents." *Royal Institute of Philosophy Supplements* 51 (2002): 233–251.
Wippel, John F. *The Metaphysical Thought of Thomas Aquinas: From Finite Being to Uncreated Being.* Washington, D.C.: Catholic University of America Press, 2000.
Wippel, John F. "Thomas Aquinas and Participation." *Studies in Philosophy and the History of Philosophy* 17 (1987): 117–158.
Wippel, John F. *The Ultimate Why Question: Why Is There Anything At All Rather Than Nothing Whatsoever?* Washington, D.C.: Catholic University of America Press, 2011.
Wood, Adam. *Thomas Aquinas on the Immateriality of the Human Intellect.* Washington, D.C.: The Catholic University of America Press, 2020.
Yourgrau, Palle. *Death and Nonexistence.* New York: Oxford University Press, 2019.
Zalta, Edward N. *Abstract Objects: An Introduction to Axiomatic Metaphysics.* Dordrecht; Lancaster: D. Reidel, 1983.
Zalta, Edward N. "Frege's Theorem and Foundations of Arithmetic." In *The Stanford Encyclopedia of Philosophy,* edited by Edward N. Zalta 2020. https://plato.stanford.edu/archives/win2020/entries/frege-theorem/.
Zoll, Patrick. "Review of Kris McDaniel, *The Fragmentation of Being.*" *Zeitschrift für Theologie und Philosophie* 143, no. 1 (2021): 145–147.

Index of Names

https://doi.org/10.1515/9783110979879-013

Index of Subjects

https://doi.org/10.1515/9783110979879-014

www.ingramcontent.com/pod-product-compliance
Ingram Content Group UK Ltd.
Pitfield, Milton Keynes, MK11 3LW, UK
UKHW041845031026
15587UKWH00003B/1019

9 783110 991307